Sustainable Agriculture *and* Organic Farming

Sustainable Agriculture *and* Organic Farming

By :

Ram Narayan Meena

Assistant Professor
Department of Agronomy
Institute of Agricultural Sciences
Banaras Hindu University, Varanasi-221005
India

SATISH SERIAL PUBLISHING HOUSE

403, Express Tower, Commercial Complex, Azadpur, Delhi-110033 (India)
Phone : 011-27672852, Fax : 91-11-27672046
E-mail :info@satishserial.com, hkjain1975@yahoo.com
Website : www.satishserial.com

Published by :

SATISH SERIAL PUBLISHING HOUSE

403, Express Tower, Commercial Complex, Azadpur, Delhi-110033 (INDIA)
Phone : 011-27672852 Fax : 91-11-27672046
E-mail : info@satishserial.com, hkjain1975@yahoo.com

ISBN 978-93-85055-76-8

Composed, Designed & Printed in India

Preface

With a view to cover up one modern **Sustainable Agriculture and Organic Farming,** I have started my sincere effort for past few years. I have tried to make it as up-to-date as possible. I am also aware of the fact that in the first edition of this book, there must be many shortcomings and printing errors. I expect that my fellow colleague teachers and specialists will co-operate with me by sending their valuable suggestions and criticisms which will be thankfully received and will help me in improving the next edition. In this book, Sustainable Agriculture and Organic Farming has been described with the ultimate aim of enriching the fertility status of the soil through organic resources.

The Sustainable Agriculture and Organic Farming book will be a vital tool for all organic food producers/organic farmers, agricultural scientists, botanists and besides those studying and teaching in organic Agriculture.

I thank Prof. Ravi Pratap Singh, Director, Institute of Agricultural Sciences for boosting up my morale and support to publish this book. I gratefully acknowledge the sincere effort of Prof. Avijit Sen, Head of department in preparation of this manuscript.

I am glad to record my thanks to my colleagues and friends for their continued interest.

February, 2016

(Ram Narayan Meena)

Contents

Chapter-1

Organic Farming and Sustainability

Organic farming is universally known to be a specialized form of farming practices, involving selected application of organic substrates such as manures, crop residues, green manuring crops, earthworm casts (vermiculture) etc. to enrich soil with adequate plant nutrients, and provide good soil structure and soil health with the aim of creating a sustainable form of farming system. Broadly, the aim of organic farming is: to create integrated, humane, environmentally and economically sustainable production systems, which maximize reliance on farm-derived renewable and natural resources to lend support to the management of ecological and biological processes and interactions, so as to provide acceptable levels of crop, livestock and human nutrition, protection from pests and diseases, and an appropriate return from the deployment of the human and other resources. There are several benefits of organic farming vis-à-vis conventional farming systems. Organic farming, among other things, promotes recycling of organic wastes, use of renewable energy and application of system-worthy low-cost and appropriate technologies using mostly the local farm-based resources. Organic farming has of late emerged as a potential alternative for meeting food security, maintaining soil fertility and increasing soil carbon pool. Organic farming technology and practices has evolved over past several decades in several parts of the world giving ways for new and emerging innovations; especially in the areas of eco-innovation for green growth, sustainable eco-friendly farming and of late green marketing. In this paper, an attempt has been made to cover the various emerging areas of organic farming and sustainable agriculture, including the evolution and benefits of organic farming vis-à-vis conventional agriculture, long-term sustainability and environmental quality, Linkage of Organic Farming with Green Supply-Chain and Food Value-Chain, eco-labelling and green marketing of organic farm produce, the future perspectives etc. The desirability of good governance and compliance of quality standards and promotion of organic foods by providing various incentives are highlighted in conclusion of the paper as viable policy measures by the producers as well as the governant as well as certifying agencies at large.

Organic farming is basically a specialized form of farming involving selected application of organic manures, crop residues, green manuring crops, earthworm casts (vermiculture) etc to enrich soil with adequate nutrients, and provide good soil structure and soil health with the aim of creating a sustainable form of farming system. Broadly, the aim of organic farming is: to create integrated, humane, environmentally and economically sustainable production systems, which maximize reliance on farm-derived renewable resources and the management of ecological and biological processes and interactions, so as to provide acceptable levels of crop, livestock and human nutrition, protection from pests and disease, and an appropriate return to the human and other resources. There are several benefits of organic farming vis-à-vis conventional farming systems. Organic farming, among other things promotes recycling of organic wastes, use of renewable energy and application of system-worthy low-cost and appropriate technologies using local resources. Organic farming has emerged as a potential alternative for meeting food demand, maintaining soil fertility and increasing soil carbon pool. Organic farming is not only about managing the soil - plant - environmental interaction in a holistic manner – it also has food quality, human health, animal welfare and socio -economic aims. As a result of these principles and philosophies, organic food has a strong brand image in the eyes of the health-, environment- and socially-conscious consumer.

Organic Agriculture – a predominant component of organic farming is a specific type of low external input agriculture that adheres to certain principles in the production and transformation of agricultural commodities. Organic agriculture may be either certified or non-certified. Certified organic agriculture must meet certain standards in the production, processing and handling of organic products. These standards are developed in accordance with basic standards established by the International Federation of Organic

Agriculture Movements. The basic standards provide a framework within which certification bodies worldwide develop their own certification standards, which may vary across countries depending upon specific local conditions. Organic standards have been established by most industrialized nations, and organic standards are specified in guidelines of the *Codex Alimentarius* of the Food and Agriculture Organization of the United Nations and the World Health Organization. Organic agriculture is based on minimizing the use of external inputs and avoiding the use of synthetic fertilizers and pesticides.

Organic agriculture has come a long way from a tiny, disorganized, idealistic fringe movement. Twenty-five years ago some may have thought it was a relic of the 1960s that was headed for the compost pile. Now most food stores carry some organic products, it is the fastest growing sector in the grocery business and the giants in the food processing and retail business, from General Mills to Wal-Mart, wants to cash in on the organic market. But rather than rejoicing at the success of the organic paradigm, some former organic supporters are asking "Can organic agriculture remain sustainable with these changes in the market?" To explore the plausible answers to this query we should be made aware and be reminded of the evolutionary processes and development of organic farming technologically, environmentally as well as from the wider economic and sustainability perspectives

Over the past couple of centuries the principles and practices of organic farming has undergone evolutionary changes; and in the process of such transitions from the conventional farming systems met with severable challenges. The various aspects of organic farming are presently engaging attention to the practitioners of organic farmers; and these include: clarity on the basic concepts and historic evolution of organic farming; aims of organic production and processing, principles of organic farming, major components and technology of organic farming. The emerging areas of organic farming are those of assessing and exploring organic products' new market potential and growth perspectives; environmental, ecological and economic dimensions of organic farming, preservation of biodiversity, soil microbial growth to leverage positive *'Rhizosphere effect'*, mode of improvement of soil health and soil fertility regime, increased usage of microbial bio-fertilisers in organic farming.

For many years the organic movement was dominated by small farms and businesses, and in most cases farmers received significant premiums for organic products. So in practice, organic agriculture contributed to the economic sustainability of farms and food businesses. Because these economic gains allowed small and medium-sized farms to thrive where similar businesses in conventional agriculture were struggling economically, organic agriculture also had social benefits for rural communities. Many organic supporters saw these benefits as equal in importance to the ecological benefits and central to the overall sustainability of organic agriculture. In recent years, however, large traditional food processors and retailers have bought out most small independent organic processing, distribution, and retail businesses. Most large processing companies and retailers are committed to minimizing the price they pay for organic products and often prefer the ease of dealing with a few large suppliers rather than many small farms.

Fortunately, as organics have become more mainstream; more farmers are learning about the environmental and economic impact of their conventional methods. Some are even making the switch to organic farming. In the U.S., depending on whose definition is used, crops that are raised without using synthetic pesticides, synthetic fertilizers or sewage sludge fertilizer – and which have not been altered by genetic engineering – are generally referred to as "organic crops." Similarly, organic animal products come from animals that have been fed 100% organic feed and raised without the use of growth hormones or antibiotics in an environment where they have access to the outdoors. While there are variations between what qualifies as organic from country to country, the concepts are fairly standard. According to the USDA, organic farms and processors:

- Preserve natural resources and biodiversity.
- Support animal health and welfare.
- Provide access to the outdoors so that animals can exercise their natural behaviors.
- Only use approved materials.
- Do not use genetically modified ingredients.
- Receive annual onsite inspections.
- Separate organic food from nonorganic food.

Benefits of Organic Agriculture

Organic agriculture can contribute to meaningful socio-economic and ecologically sustainable development, especially in poorer countries. This is due on the one hand to the application of organic principles, which means efficient management of local resources (e.g. local seed varieties, manure, etc.) and therefore cost-effectiveness. Various organic farming technologies have been utilized for several years world over to allow agriculture and allied farming activities to extract and judiciously utilize natural resources, while conserving soil, water, energy, and biological resources; sans, of course, the detriment to ecology and environment. Broadly, the potential benefits of organic farming technologies are higher soil organic matter, organic carbon and nitrogen, lower fossil-fuel based energy inputs, comparable yield performance as are quite similar to those of conventional systems. The other incidental benefits are conservation of soil moisture and water resources (especially advantageous under drought conditions). Experiences of traditional farmers have shown that even the conventional agriculture can be made more sustainable and ecologically sound by adopting some traditional organic farming technologies. Organic agriculture (OA) and conventional agriculture (CA) represent two polar approaches to farming, both of which employ their own methods and farming practices. OA and CA can both hold their own challenges and implications within the global food-chain, particularly with respect to the impending global food crisis, coupled with climate change consequential to global warming. These issues present pressing agricultural questions leading us to ask how our farming methods will adapt to feed the world's mammoth population while maintaining the viability and health of the world's ecosystems.

Organic agriculture shows several benefits, as it reduces many of the environmental impacts of conventional agriculture, it can increase productivity in small farmers' fields, and it reduces reliance on costly external inputs, and guarantees price premiums for organic products. Organic farmers also benefit from organizing in farmer cooperatives and the building of social networks, which provide them with better access to training, credit and health services. Organic agriculture generally reduces the vulnerability of farmers as the higher organic prices act as buffer against the low prices and price volatility of conventional markets, as organic systems are often more resilient against extreme weather events, and as the often diverse organic crop-livestock systems provide a diverse set of outputs.

Systematic research studies are now being conducted to evaluate relative environmental vis-à-vis economic benefits of organic farming as opposed to conventional farming approach. Some of the current endeavours put up by the lead research establishments in this direction encompass review of experiences of different agrochemical industries in organic and conventional farming systems, measurements of profitability of organics-based alternate farming systems, analysis of socio-economic approach of sustainable farming systems vs. conventional agriculture, organic farming for food production stability, climate mitigation and adaptation actions, organic-based integrated farming systems to help build synergy between conventional and organic farming.

The green marketing is often considered a solution to the many issues of the degradation of the environment and the consumers' health. The ecological, green or environmental marketing attempts to connect the classical components of the marketing and management to the ecological issues. The key-concept attempting to define the green marketing is how the responsibility and environmental issues are integrated into the concept of (the) marketing management. In this respect, the green marketing might be defines as the holistic management process responsible for identifying, anticipating and satisfying the requirements of the customers and society, in a profitable and sustainable way. Green marketing incorporates a board range of activities about ecological products, including organic products as an important part, about changes in the production process, in the packaging as well as modifying promotion and distribution.

Preserving Long-Term Sustainability and Environmental Quality

In recent years, many studies have shown that integrated farming system, combining organic and conventional farming system could be a better option to ensure long-term sustainability without sacrificing profitability but some sacrifice to the environmental quality. Hence, organic agriculture - the predominant component of organic farming, is increasingly being seen to be a sustainable way of farming without chemical inputs during cultivation, whereas integrated farming system is a sustainable way of farming which falls somewhere in between the conventional and the organic farming system. Organic and integrated agriculture are the sustainable farming systems that have been developing noticeably during the last decade. Organic Farming (OF) is one of the agro-ecological approaches needed to grow enough food for the increasing population. This approach minimizes external inputs such as chemical fertilizers, pesticides to produce non-toxic crops. Thus, it is less environmentally damaging and has much potential to produce more food in a sustainable manner.

While conventional farming systems face serious problems of sustainability, organic agriculture is seen as a more environmentally-friendly system since it favours renewable resources, recycles nutrients, uses the environment's own systems for controlling pests and diseases, sustains ecosystems, protects soils, and reduces pollution. At the same time, organic farming promotes animal welfare, the use of natural foodstuffs, product diversity and the avoidance of waste, among other practices. However, the future of organic agriculture will depend on its economic viability and on the determination shown by governments to protect these practices.

Agricultural researchers widely recognise the importance of sustainable agricultural production systems and the implicit need to develop appropriate methods to measure sustainability. An integrated farming system consists of a range of resource-saving practices that aim to achieve acceptable profits and high and sustained production levels, while minimizing the negative effects of intensive farming and preserving the environment. Based on the principle of enhancing natural biological processes above and below the ground, the integrated system represents a winning combination that (a) reduces erosion; (b) increases crop yields, soil biological activity and nutrient recycling; (c) intensifies land use, improving profits; and (d) can therefore help reduce poverty and malnutrition and strengthen environmental sustainability. An integrated farming system involving such components as crops, forestry, livestock, fishery, piggery, poultry, apiculture, pisciculture/aquaculture, vermiculture, organic waste recycling, renewable energy deployment to farming system etc. can make a significant synergistic contribution to sustainable development of farming system on a long-term basis.

Organic agriculture can contribute to meaningful socio-economic and ecologically sustainable development, especially in poorer countries. This is due on the one hand to the application of organic principles, which means efficient management of local resources (e.g. local seed varieties, manure, etc.) and therefore cost effectiveness. On the other hand, the market for organic products - at local and international level - has tremendous growth prospects and offers creative producers and exporters' excellent opportunities to improve their income and living conditions. Organic agriculture reduces the risk of yield failure, stabilizes returns and improves the quality of life of small farmers' families.

Organic Farming with Green Supply-Chain

Organic farming is now being linked to the *green supply-chain* in one hand; the *food value-chain* on the other. Food value-chains represent a business model in which producers and buyers of agricultural products form strategic alliances with other supply chain actors, such as aggregators, processors, distributors, retailers, and consumers, to enhance financial returns through product differentiation that advances social or environmental values. Partners in these business alliances recognize that creating maximum value for their products depends on interdependence, collaboration, and mutual support. The food value-chain model is gaining traction because it responds to agricultural and food industry consolidation that has placed intense market pressure on small, marginal and mid-sized farmers.

Central to the notion of food value-chains is the idea that transparent and trusting relationships between supply-chain partners can produce positive, 'win-win' outcomes for all parties. In this model, consumers, farmers, distributors, and others in the chain of food business activity, from planning and planting to processing and selling, see results and reap rewards. The gains of producers are not achieved at the expense of distributors or retailers, or vice versa, because the structure of food value chain transactions facilitates the sale of a broader range of well-differentiated food products, priced to reflect the incorporation of both social and private benefits, which are more closely tailored to the preferences of specific consumer segments. The *organic principles of farming* based of integration of health, ecology, stakeholders partnership, fairness and care (i.e. value-chain), should, ideally, include these multiple functions.

Agri-food systems are undergoing rapid transformations and the emergence of integrated food supply chains is one of the most visible market phenomena in India. Increasing concentration on processing, trading, marketing and retailing is being observed in all the segments of supply chains. Organization of agriculture, along the *value-chain framework* has been conceived as one of the strategies to bring more efficiency in the agricultural sector. The *value-chain network* may now be defined as a range of activities that are required to bring a product from its conception, through its designing, sourcing of raw materials and intermediate inputs, marketing and distribution, to the final consumer. There has been an increasing emphasis on the development of efficient organic value-chains in India and several innovative and successful value-chains have emerged to popularize organic farming among the farming communities in order to exploit the full market potential of organic produce in future.

An ideal value chain should bring all the stakeholders engaged in the production system on a common platform to contribute their best, while ensuring fair deal and transparency. The value chain will include all the input suppliers, technology delivering agencies, scientists indirectly engaged in developing appropriate technologies and extension officers who are involved in capacity building and providing various services to farmers. The stakeholders involved in post-production activities are the agencies organising collection, grading, storage, transportation, processing and marketing of the produce. Agencies like financial institutions and market information centres are also part of the value chain. Efficient linkage of various stakeholders improves production, price realisation and profitability. Research, training, capacity building and continuous innovation in developing organic farming models and testing their economic viability and adaptability will add to the organic value-chain in a sustainable manner.

The increasing pace of globalization and industrialization has revolutionarised the way the corporate world is responding the changed life-style with time. Some of the major casualties of the energy-intensive mode of industrialization are enhanced rate of greenhouse gas emissions, global warming and ecological degradation. As the public becomes more aware of environmental issues and global warming, consumers are asking more questions about the eco-friendliness of products they are purchasing. Companies are confronted with embarrassing questions like how green their manufacturing processes and supply chain are, the state of their carbon footprints and how they recycle and dispose wastes etc. Consumers increasingly prefer to purchase products that are free of toxins, produced with minimum level of pollution-linked contaminants and with minimal environmental impact. Increasing evidences are accumulating that a gulf of difference exists between consumers' green claims and the certified quality of 'green' products. Companies that successfully adopt a 'green' policy can generate profits, provide positive social impact, and reduce environmental impact as there is an apparent link between improved environmental performance and financial gains.

Realizing this perspective, companies have looked to their supply chain and explored areas where can expect more profits. Prominent features of leading green supply chains include an emphasis on life-cycle costing, asset efficiency, and waste reduction and service innovation and recycling. Executed effectively, Green Supply Chain Management (GSCM) stimulates product and service innovation, improves asset utilization, and deepens customer relationships and service levels through a shared focus on reducing waste and cost. Several companies have gone one step further by e green supply chain management (GSCM) as a competitive advantage. The premise of GSCM is that sharing and integrating environmental ideas and concerns across organizational boundaries will greatly enhance green manufacturing. More manufacturing firms have recognized the importance of GSCM practices and have begun to foster long-term partnerships with their suppliers to increase environmental performance. Green Supply Chain Management (GSCM) improves Green marketing operations by employing an environmental solution in the following ways:

- Improves agility - GSCM helps to mitigate risk and speeds up innovations;
- Increases adaptability - green supply chain analysis often leads to innovative processes and continuous improvements;
- Promotes alignment - GSCM involves negotiating policies with suppliers and customers, which results in better alignment of business processes and principles.
- Core focus of the lectureship will be on the necessity of 'green' and on green operations (network design and reverse logistics, transportation, green manufacturing and re-manufacturing and waste management).

Organic Certification, Carbon Foot-printing, Eco-labelling and Green marketing

The term '*carbon footprint*' is commonly used to describe the total amount of Carbon-di-Oxide and other greenhouse gas (GHG) emissions for which an individual or organization is responsible. Footprints can also be calculated for events or products. Certification marks, labels and logos are increasingly being used by brand owners to signal their green credentials and so boost their market share. A properly controlled *eco-label* offers consumers a guarantee that a product or service has been independently verified to meet given environmental standards. In Australia, for example, the Greenhouse Friendly™ label is a registered certification mark, administered by the Government Department of Climate Change. Some companies are developing their own eco-standards and product labelling. MNCs like BASF and Philips have launching their - Green Logo and tick symbol last year to identify products with significantly better energy efficiency than the nearest competitor products. Eco-labelling could be of two kinds: Performance based and Process-based. An eco-labelling scheme based on product performance, such as labels claiming biodegradability of packaging, sustainability in production processes or non-pollutant aspects of product usage will require certain trust on the part of the consumer. A systems approach to the eco-labelling of food and other products requires the involvement of multiple stakeholders as one party's actions affect others' environmental performance.

Organic certification is a certification process for producers and marketers of organic food and other organic agricultural/ non-agricultural organic products. In general, any business directly involved in food production can be certified, including seed suppliers, farmers, food processors, exporters, retailers and hotels and restaurants. Requirements vary from country to country, and generally involve a set of production standards for growing, storage, processing, packaging and shipping etc. In some countries, certification is overseen by the government, and commercial use of the term *organic* is legally restricted. Certified organic producers are also subject compliance to the same agricultural, food safety standards and other government regulations that apply equally to non-certified producers.

An eco-label, in the context of organic farming identifies a claimed to be organic product that meets a wide range of environmental performance criteria or standards. Developed by governments, manufacturers, and third-company organizations, eco-labeling is a voluntary approach to environmental certification practiced around the world. In contrast to "green" symbols or claims, an 'eco-label' is given to products that have met specific environmental safeguard criteria. As there is a wide range of products available in the market, environmental performance labels and declarations vary greatly. The growing number of environmental claims led the U.S. Federal Trade Commission (FTC) in 1992 to issue Title 16-Part 260 CFR: Guides for the Use of Environmental Marketing ("Green Guides"). The FTC issued 'Green Guides' to help marketers avoid making environmental claims that are unfair or deceptive under Section 5 of the FTC Act. This includes guidance on claims for biodegradable, compostable, recyclable, recycled, and ozone-safe content. The FTC has updated the guides and is currently undergoing an extensive review process. General benefits of eco-labelling include: economic incentives for better long-term custodianship and availability of natural resources vital for economic welfare; competitive and comparative advantage of export products through product differentiation which is realised through price premiums, long-term contracts and market access; a platform for innovation with the use of more environmentally friendly products such as lighting or refrigeration, with knock on benefits in other parts of the economy; and last but not the least providing assistance to countries to fulfil commitments made under environmental agreements such as biodiversity.

Some recent estimates suggest that there are over 400 existing eco-labels marking consumer products in nearly every category and the number are growing rapidly. In the wake of the recent global recession and major economic crisis, however, it appears that consumers are less motivated to purchase green products. For example, in the United Kingdom, sales of certified-organic fruits, vegetables, and meat have plummeted 12.9% in the past year. In other words, green claims are proliferating fast in the marketplace, but there are serious questions about their efficacy in driving sustainability outcomes and their success in creating real consumer preference.

The concept of labeling organic products has been in existence for quite some times now. At present, there are over 300 eco-labels, according to cataloguer Ecolabelindex.com (2010). Competition between eco-labels carries benefits vis-a-vis pitfalls; it can raise the bar on performance, but it also tends to create confusion among consumers, who are left wondering whether organic products are genuine, eco-friendly, recycled or recyclable. To-date, there has been limited research on eco-label design, and very little is known about what drivers affect a label's market penetration and associated product sales.

Certification for Green Food production involves the regulation of inputs, with the objective of reduced use of pesticides and other agro-chemicals, the oversight of production, and the residue testing of the claimed organic produce. In 1995, Green Food certification is usually split into 'Grade-A' and 'Grade-AA'. It is this bifurcation of Green Food standards that laid the groundwork for the rapid articulation from 'Green Food certification to Organic certification'. China, for example, has over 30 million hectares of eco-food production. China's total eco-labelled food production area is 28% of China's total of 122 million hectares of agricultural land.

Eco-labels have emerged as one of the main tools of green marketing. Although a great deal of effort has gone into in making them more effective and efficient, the market share of eco-labelled products is still low, partly because they have been addressed mainly to 'green' consumers. Eco-labels are intended as a means for consumers to make choices that will reduce environmental impact and enable them to influence how products are made. In the Nordic Countries, there are eco-labels for 55 product groups and 2800 products. In Japan, 64 product groups have criteria established for eco-labels and more than 5000 products have been accepted.

The Food Alliance is a non-profit organization dedicated to marketing sustainable agriculture. They have a certification program for farms that:

- Provide Safe and Fair Working Conditions
- Ensure the Health and Humane Treatment of Animals
- Do Not Use Hormone or Antibiotic Supplements
- Do Not Raise Genetically Modified Crops or Livestock (GMOs)
- Reduce Pesticide Use and Toxicity
- Protect Water Resources
- Protect and Enhance Soil Resources
- Provide Wildlife habitat
- Continually Improve Practices.

While their regulations on pesticide and fertilizer use are less stringent than organic standards, their economic and animal welfare requirements are more rigorous.

Certification and the standardizing eco-labelling procedures are the basic foundations for green marketing of organically produced eco-friendly products; also called 'green products' or 'eco-products'. Certification and labeling systems serve as tools to enhance distribution and market development, create trust, and foster confidence. It is a commitment from producers/farmers to work with certain standards of production. In 2009, the Canadian government implemented the Organic Products Regulation to regulate organic certification. The number of certified organic producers for the local market is growing and there are now Participatory Guarantee Systems (PGS) initiatives on all continents in terms of the number of farmers involved, with Latin America and India being the leaders. However consumers' confidence in certification standards in other countries and trust in their labels and products could be increased by the consolidation of standards and regulations between countries like Canada and the US, the world's first fully reciprocal agreement between regulated organic systems.

The latest developments in organic farming and consumption pattern show dynamic market growth; however the uptake of organic farming by farmers is seen to be fairly slow. In recent years, global market growth has grown over 5%, while the agriculture area in general and organic sector in particular, and have been rather sluggish. Bottlenecks in the adoption of organic farming are often due to production and/or processing (including post-harvest management) problems or even the fact that farmers are not always convinced that organic methods can solve farming problems, such as fertilization, plant protection, animal health, efficient use of workforce, marketing diversity, overall cost effectiveness and competitive economic advantage. Although these problems can sometimes be resolved through continuous learning on existing and regionally practiced methods; innovations nonetheless, are imperative to making organic farming more adaptive, sustainable and regenerative for fetching the benefits of organic farming and organic produce to a wider consumer-base. Farmers must therefore continuously adopt production and management systems in order to maintain and enhance the competitiveness and sustainability of their businesses. The development and implementation of innovations require both in terms of information and the farmers' willingness to change daily work schedules and methods. Learning and knowledge transfer among farmers, technology developers, experts and university teams ensure the development and application of innovative ideas which are crucial for a sustainable growth in food (and non-food) production.

The ecological significance of sustainable development has increased dramatically since the so called 'green economy' and 'organic production' of 'eco-friendly organic produce' have been turning into sources of competitive advantage for the country of origin in the international markets. A high number of new or modified products (including genetically modified products) are often combined with process innovations; in addition to product innovation. Three very innovative fields in the organic food industry are in the area of 'genetically modified organisms (GMO)', 'functional food' and 'organic food', representing market opportunities, which, however, are also experiencing different obstacles and drivers. A future innovation challenge to the agro-food industry, with focus on organics, will be to tackle the multiple new scientific approaches and technical opportunities that have an interdisciplinary character. The creation and building-up of interfacing competencies as well as the establishment of new external knowledge and competence networks seems to be of strategic relevance for future growth of these categories of industries. The target for policy should be to support the advances of the knowledge base of the food industry companies themselves and the diffusion of new scientific approaches and technology, and not solely concentrate on stimulating knowledge generation with relevance for the food industry in general; and life-style organic products (eco-products in particular).

Organic Food Market and Future Potential

According to the 2013 statistics published by Research Institute of Organic Agriculture and International Federation of Organic Agricul-ture Movement (IFAOM, 2013), the countries with the largest organic markets in 2012 were the U.S., Germany and France. These statistics further state that the largest single market was the U.S. (approximately 44% of the global market), followed by the European Union (approximately 41%). The highest per-capita consumption was in Switzerland, Denmark and Luxembourg. The highest market shares were reached in Denmark, Switzerland and Austria. Other vital attributes of the organic food market include (IFAOM, 2013).

Alternative food systems: Organic food grown on large machine scale (over 500 acres), referred to as the "industrial organics," originate from undiversified farms for large corporations. These farms tend to have large fields of a single crop or to raise only livestock. They generally sell wholesale and almost never directly to the customer. Industrial organics are always certified. The customer, in almost all cases, never talks to the grower(s) from these farms. Recently, it has been observed that consumers of food from industrial organics have limited knowledge about their source of food, causing concern among consumers. These concerns are directly related to economic, environmental and social ideals such as production systems, size of the operations, distribution systems and channels and capital intensity. This trend has given rise to the popularity alternative food systems (AFS). AFS are perceived as sustain-able and economically, socially and environmentally and found to be more viable than standard systems. The slow food movement, box delivery and local farmers markets are good examples of innovative food systems that are based on low-carbon food distribution systems.

Market segmentation: The market segments for organic food are expanding and can be classified into seven categories, according to popular 2013 Chicago IRI research. However, the two shopper segments that account for about half of the natural and organic sales are termed True Believers and Enlightened Environmentalists. The former believe in the benefits of natural products for themselves and their families, and the latter care about protecting the environment. Currently, these consumers constitute a large portion of organic food buyers. But retailers and supermarkets must pay attention, since the organic grocery space is set for a dramatic growth in the food industry (NASDAC, 2013).

Future growth: According to The United States Organic Food Market Forecast & Opportunities, 2018, the greatest demand for organic food is generated from the U.S. residential sector. The growth forecast for the organic food market of the U.S. is at a compound annual growth rate of about 14% from 2013 through 2018 (Tech. Science Research, 2013). Also, organic fruits and vegetables will continue to dominate the market through 2018. As organic food market revenues expand, the demand for organic meat, fish, poultry, etc. will gain market share. Further, according to forecasts, the education and restaurant sectors will also contribute substantially to the organic food market, due to rising awareness of the benefits of organic food.

For many years the organic movement was dominated by small farms and businesses, and in most cases farmers received significant premiums for organic products. So in practice, organic agriculture contributed to the economic sustainability of farms and food businesses. Because these economic gains allowed small and medium-sized farms to thrive where similar businesses in conventional agriculture were struggling economically, organic agriculture also had social benefits for rural communities. Many organic supporters saw these benefits as equal in importance to the ecological benefits and central to the overall sustainability of organic agriculture. In recent years, however, large traditional food processors and retailers have bought out most small independent organic processing, distribution, and retail businesses. Most large processing companies and retailers are committed to minimizing the price they pay for organic products and often prefer the ease of dealing with a few large suppliers rather than many small farms.

Future prospects of organic farming should safeguard the basic interests of consumers in the domain of health and prosperity, giving importance to consumption of organic food products - a way to healthy life, voluntary food-labelling to meet lifestyle requirements, exploration of new business opportunities by promoting specialty food products (organics), ensuring fairness to growing green-conscious customers, protection of ecology, designing organic research agenda for a sustainable future, innovations in organic gardening for bestowing a new age-defying lifestyle product concept.

Keeping in view the growing importance currently being given by the government and the implicit desirability of embracing eco-friendly and pollution-free cultivation and farming practices, organic farming seems to be a long-term pragmatic solution. With the increasing expressed concerns and consciousness for conservation of natural resources, protection of biodiversity of flora and fauna, protection of environmental quality, improving health and hygiene with changing lifestyle throughout the world organic Farming and Sustainable Agriculture has a great future.

In India, the relative lack of national rules, regulations and specific standards relating to organic food production, inadequate certifying agencies and unrecognized 'green' marketing and retailing channels have not only been confusing for producers and consumers alike, but have prevented farmers from exploiting the export market advantages of organic production. This is a major missed opportunity because most small and marginal farmers in India have actually been practicing organic farming as part of traditional cultivation practice. Thus they have used local or own-farm derived renewable resources and managing self-regulating ecological and biological processes. In fact, this is usually found to be absolutely necessary simply in order to cultivate acceptable levels of crop, livestock and human nutrition products while protecting them from pests and diseases through bio-chemicals and bio-fertilizers (such as Neem extract). However, it is true that the higher cost of such inputs and processes compared to industrially generated fertilizers and pesticides has encouraged many farmers to shift production patterns.

No system of food production can guarantee perfect sustainability. The organic standards have flaws and gaps, but there are a variety of efforts to address at least some of them. In addition, consumers, farmers, and processors can seek to buy and produce food with other ethical qualities, such as economic justice (fair trade) and local food. As yet, they account for a small percentage of the food sold both in the US and around the world, but organic and other value-based foods are a rapidly growing part of the market. Questions about the environmental and social sustainability of our food supply are becoming mainstream. However, the complexity of the agro-ecosystem and food system make reform challenging.

□□□

Chapter-2

Integrated Farming System Approach

Integration of various farm enterprises in a farm ensures growth and stability in overall productivity and profitability. It also ensures recycling of residues, optimization of resources, minimization of risk and generation of employment. Various enterprises that could be included in farming system are crops, vegetables, fruits, flower cultivation, dairy, poultry, fish, goat, pig, sericulture, mushroom cultivation, agro-forestry, bee keeping, silviculture, agro-based industries and food processing. A judicious mix of enterprises complementary to cropping and suited to the given farm situation and farmer's preference would bring overall prosperity.

Although India is currently self-sufficient in cereals, still millions of people are inadequately fed, as economic access to food could not be achieved due to poor purchasing power of the farmers. The per capita availability of land decreased from 0.5 ha in 1950-51 to 0.15 ha in 1999-2000, and is further likely to be reduced to 0.08 ha by 2020. There is hardly any scope for horizontal expansion of land. The situation has further aggravated by following the Uruguay Agreement and Globalization of Markets which calls for competitiveness and efficiency in agricultural production, renders the agricultural sector to face challenges for ecological balance, social justice, energy and employment generation. Since majority of farmers in India are resource poor and have marginal and small economic holding, it is very difficult to enhance the farm family income unless crop based agriculture is supplemented by some other farm enterprises like livestock production, bee keeping, poultry, fisheries, horticulture etc. Being inter and multidisciplinary approach, farming system would be highly effective in providing balanced food, regular employment, sustaining soil health, mitigating aberrant weather situation, and increased the farm productivity and farm family income which ultimately increases the purchasing power of farmers. Thus, the future agricultural will be guided by not only the compulsions of food and nutritional security, but also by the concerns of environmental protection, sustainable soil health as well as profitability of farm enterprises.

The integrated system approach is not only a reliable way of obtaining fairly high productivity with substantial fertilizers economy, but also a concept of ecological soundness leading to sustainable agriculture.

Farming system as an approach to agricultural research and development that views the whole farm as a system and focuses :The interdecencies between the component under the control of members of the house hold and Interact of these components with physical, biological and socio-economic factors which are not under the control of households.

Farming system is a resource management strategy to achieve economic and sustained production to meet diverse requirement of farm household while preserving resource base and maintaining a high level environmental quality.

Farming system is complex inter-related matrix of soil, plants animals, implements, power, labour, capital and other inputs controlled in part by farm families and influenced by varying degree of political, economic, institutional and social forces that operate at many levels. Under farming system farm is viewed in holistic manner. Farmers are subjected to many socio-economic biophysical institutional, administrative and technological constraints. Farming system conceptually is a set of elements or components that are inter related which interact among themselves. At the center of the interaction is the farmer exercising control and choice regarding the type and results of interaction.

Integrated Farming is defined as biologically integrated farming system which integrates natural resources and regulation mechanisms into farming activities to achieve maximum replacement of off-farm inputs; secures sustainable production of high quality food and other income; eliminates or reduces sources of present environment pollutions generated by agriculture and sustains the multiple function of agriculture (IOBC, 1993).

Scope of Farming System

Farming enterprises include crop, livestock, poultry, fish, tree crops, plantation crops, sericulture, etc. A combination of one or more enterprises with cropping when carefully chosen, planned and executed gives greater dividends than a single enterprise, especially for small and marginal farmers. Farm as a unit is to be combined with crop production activity. Integration of farm enterprises depends on many factors such as :

- Soil and climatic features of the selected areas
- Availability of the resources, land, labour and capital
- Present level of utilization of resources
- Economics of proposed integrated farming system
- Managerial skill of the farmer.

Farming System Concept

Sustainability is the objectives of the farming system where production process is optimized through efficient utilization of inputs without infringing on the quality of environment with which it interacts. The specific objectives are :

- To identify existing farming system in specific areas and assess their relative viability.
- To formulate farming system models involving main and allied enterprises for different farming situations.
- To ensure optimal utilization and conservation of available resources and effective recycling of farm residues within a system and
- To maintain sustainable production system without damaging resources/environment.

As a whole the concept of farming system research is to integrate farm enterprises for optimum utilization of resources, minimize environmental pollution, increase farmer's income, create regular employment opportunities throughout the year, increase export potential, support rural agro-industry and ultimately food security.

Role of Farming System

- Food security
- Provides balanced food
- Quality food basket
- Higher productivity and enhanced farm income
- Employment generation
- Poverty alleviation
- Social upliftment
- Effective recycling of resources
- Sustaining soil health
- Conservation of natural resources
- Reducing risk factor
- Minimize environmental pollution

Constraints and Opportunities

- Lack of appropriate technology
- Inadequate training
- Lack of rural infrastructure
- Lack of farmer's participatory research
- Inadvertent avoidance of farm women
- Inadequate institutional support
- Socio-economic constraints
- Policy implication

Future Research Thrust

- Development of location specific farming system models based on farmers participatory programme is need of the time for productivity enhancement and also for the improvement in the livelihood of the farming community.

The modern production technologies with high input requirement offers tangible solution to the small and medium farmers problem of low income and low productivity from cropping. To meet the growing demands, and to solve small and medium farmer's problems and to achieve the targeted production levels with stability, ecological sustainability and equitability in the 21st century, a viable technology called integrated farming system approach suited to varied farm situations and agroclimatic conditions seem to be a potential tool.

Terminology to animal component in IFS

The integrated farming system approach is not only a reliable way of obtaining fairly high productivity with substantial fertilizer economy, but also a concept ecological soundness leading to sustainable agricultural.

Buller or Nymphomniac	A cow apparently always in heat
Cryptorchid	A male animal in which one or both the testicles fail to descend into the scrotal sac. This defect is heritable and hence cryptorchids should not be used for breeding.
Free-martin	When twin calves of different sexes are born, the bull calf is usually sexually normal, but often the heifer calf is sterile and the external genitalia are abnormal in structure. Such female calves are known as free-martins.
Crone	An old broken-mouthed ewe which has been retained in a breeding flock beyond the normal time because of her excellent breeding performance.
Gimmer	A female sheep between first and second shearing.
Seggy	A ram castrated after servive.
Open animal (gilt, cow, sow or ewe)	Female animals that have not been bred.
Card, Crit or Runt	The smallest pigling in a litter, usually the last to be farrowed.

Definition of Common Terms Pertaining to Livestock

Details	Cattle	Buffalo	Sheep	Goat	Pig
Species called as	Bovine	Bovine or Butaline Herd	Ovine	Caprine	Swine or Sus
Group of animals	Herd	Herd	Flock	Flock or Band	Drove or Stock or Herd
Adult female	Cow	She buffalo or buffalo cow	Ewe	Doe	Sow
Adult male	Bull	Buffalo bull	Ram or tup	Buck	Boar
Young male	Bull calf	Buffalo bull calf	Ram lamb or Tup lamb	Buckling or male kid	Boarling
Young female	Heifer calf	Buffalo heifer calf	Ewe lamb or Gimmer lamb	Goatling	Gilt
Newborns	Calf	Buffalo calf	Lamb	Kid	Piglet or Pigling
Castrated male	Bullock or Steer	Buffalo bullock	Wether or Wedder	Castrated	Hog or Stag or Barrow
Castrated female	Spayed	Spayed	Spayed	Spayed	Spayed
Female with its offsprings	Calf at foot	Calf at foot	Suckling	Suckling	Suckling
Act of parturition	Calving	Calving	Lambing	Kidding	Farrowing
Act of Mating	Serving	Serving	Tupping	Serving	Coupling

Recorded Lifespan of Various Animal Species

Species	Average years	Maximum years
Livestock		
Cattle	20-25	30
Goats	8-10	18
Horses	20-30	50
Sheep	10-15	20
Swine	16	27
Pets		
Cats	13-17	21
Dogs	13-17	34
Guinea pigs	2 plus	6 plus
Hamsters	1	2-3
Mice	1-2	3 plus
Rabbits	5-6	13
Rats	2-3	4
Monkey (rhesus)	15	29
Poultry		
Chichen	4	30
Geese	30	50
Turkey	-	12 ½
Pet birds		
Canary	-	24
Lovebirds	-	8 plus
Parrots (various species)	-	30-65
Others		
Snakes (various species)	-	6-20

Features of the Reproductive Cycle (Pets)

	Cats	**Dogs**	**Guinea pigs**	**Hamsters**	**Mice**	**Rabbits**
Age at puberty	6-15 months	6-12 months	55-70 days	5-8 weeks	35 days	5½ - 8½ months
Cycle type	Provoked ovulation, seasonally polyestrous, spring and early fall	Monstrous, all year, but mostly late winter and summer	Polyestrous	Polyestrous	Polyestrous	Polyestrous, induced ovulation
Cycle Length	15-21 days	6-7 months	16 days	4 days	4 days	1 month or more
Duration of heat	9-10 days in absence of male, 4-6 days if mated	4-14 days standing heat	6-11 hours	10-20 hours	9-20 hours	1 month or more
Best time for breeding	Daily from 2 day of heat	One alternate days from day 2 to end of heat	10 hours after start of heat	At start of heat, 8 -10 p.m.	At start of heat	Any time
First heat after birth	4-6 weeks	3-5 months	6-8 hours	1-2 weeks after litter removed	2-4 days after litter removed	Immediate
Number of young	1-10	1-22	1-8	1-12	1-12	3-13

Common Names for the Sexes, Young, Group and Birthing of Various Animals

Animal	Male	Female	Young	Group	Name for giving birth
Antelope	Buck	Doe	Kid	Herd	Kidding
Bear	Boar	Sow	Cub	Sleuth	Cubbing
Beaver	Boar	Sow	Pup	Colony	-
Bird	Cock/ stag	Hen	Fledgling/ nestling	Flock	Hatch
Bison	Bull	Cow	Calf	Herd	Calve
Bobcat	Tom	Lioness/ queen	Kitten	Litter	-
Cat	Tom	Pussy/ queen	Kitten	Clowder	Queening
Cattle	Bull	Cow/heifer	Calf	Herd/drove	Calve
Chicken	Rooster/ cock	Hen/pullet	Chick	Flock	Hatch
Deer	Buck/ stag	Doe	Fawn	Herd	-
Dog	Dog	Bitch	Pup/ puppy	Kennel	Whelp
Donkey	Jackass	Jennet/ Jenney ass	Colt	Herd	Foal
Duck	Drake	Duck	Duckling	Flock	Hatch
Elephant	Bull	Cow	Calf	Herd	Calve
Fox	Reynard	Vixen	Cub/pup	Earth/skulk	Pupping
Giraffe	Bull	Cow	Calf	Herd	Calve
Goat	Billy/ buck	Nanny	Kid	Trip	Kidding
Goose	Gander	Goose	Gosling	Gaggle/flock	Hatch
Hog	Boar	Sow/gilt	Piglet/ shoat	Herd/drove	Farrow
Horse	Stallion /stud	Mare/dam	Foal colt (male) filly (female)	Stable/herd	Foal
Kangaroo	Buck/ boomer	Doe/llyer	Joey	Troop/herd	-
Lion	Lion/ tom	Lioness/ she lion	Cub	Pride/flock	Whelp

Ostrich	Cock	Hen	Chick	Flock	Hatch
Owl	Owl	Jenny/ howlet	Howlet/ owlet	Flock	-
Ox	Steer	Cow	Stot	Herd/drove	Calve
Rabbit	Buck	Doe	Kitten	Colony	-
Rat	Buck	Doe	-	Colony	-
Seal	Bull	Cow	Pup	Herd/harem /rookery	-
Sheep	Buck/ ram	Ewe/dam	Lamb/ lambkin	Flock/ hurtle	Lamb
Turkey	Tom	Hen	Poult	Flock	Hatch
Walrus	Bull	Cow	Cub	herd	-
Whale	Bull	Cow	Calf/cub /pup	Herd/pod	Pupping
Wolf	He-wolf	She-wolf	Cub/pup	Pack	Whelp
Zebra	Stallion	Mare	Colt	Herd	Foal

Gestation Periods

Species Livestock	**Time (days)**	**Species Wild animals**	**Time (days)**	**(Months)**
Ass	365	Ape, Barbary	210	
Cattle		Bear, black		7
Angus	281	Bison		9
Ayrshire	279	Camel	410	
Brown Swiss	290	Coyote	60-64	
Charolais	289	Deer, Virginia	197-220	
Guernsey	283	Elephant		20-22
Hereford	285	Elk, Wapiti		8½
Holstein	279	Giraffe		14-15
Jersey	279	Hare	88	
Red Poll	285	Hippopotamus	225-250	
Shorthorn	282	Kangaroo, red	32-34	
Indian	282	Leopard	92-95	
Goat	148-156	Lion	108	

Horse, heavy	333-345	Llama		11
Horse, light	330-337	Marmoset	140-150	
Pig	112-120	Monkey, macaque	150-180	
Sheep		Moose		
Mutton breeds	144-147	Musk ox		9
Wool breeds	148-151	Opossum	12-13	
Pets		Porcupine	112	
Cat	59-68	Pronghorn	230-240	
Dog	56-68	Raccoon		7-8
Guinea pig	58-75	Reindeer		
Hamster	15-18	Rhinoceros		
Mouse	19-31	African	530-550	
Rabbit	30-35	Seal		11
		Shrew	20	
Fur Animals		Skunk		62-65
Chinchilla	105-115	Squirrel, grey	44	
Ferret	42	Tapir	390-400	
Fisher	338-358	Tiger	105-113	
Fox	49-55	Walrus		12
Marten, European		236-274	Whale, sperm	16
Pine Marten	220-265	Wolf	60-63	
Mink	40-74	Woodchuck	31-32	
Muskrat	28-30			
Nutria (coypu)	120-134			
Other	270-300			

FAO Statistics on Livestock Population

The table below is latest FAO Statistics on Livestock of the countries in Asia-Pacific Region, 1988.

Country	Cattle	Buffalo	Pig	Sheep	Goat	Chicken	Duck
Bangladesh	23,500	1,964	-	1,123	10,700	64,000	32,000
Bhutan	409	7	63	27	32	-	-
China	73,946	20,860	335,116	102,650	77,932	1,850,000	323,000
India	201,401	75,605	10,519	56,980	108,493	206,000	9,000
Indonesia	6,470	3,049	6,500	5,488	13,704	434,000	29,000
Iran	8,350	230	-	34,500	13,600	105,000	-
Japan	4,667	-	11,725	29	48	351,000	-
Kampuchea	1,950	700	1,500	1	1	7,000	3,000
Korea DPR	1,250	-	3,100	372	285	20,000	-
Korea REP	2,386	-	4,281	3	166	59,000	1,000
Laos	590	1,000	1,520	-	76	9,000	-
Malaysia	636	245	2,258	102	347	58,000	4,000
Mongolia	2,600	-	140	13,200	4,450	-	-
Myanmar (Burma)	10,000	2,200	3,000	295	1,100	34,000	6,000
Nepal	6,374	2,890	479	833	5,125	10,000	-
Pakistan	17,300	13,900	-	27,000	32,300	152,000	1,000
Philippines	1,600	2,890	7,580	30	2,130	60,000	5,000
Sri Lanka	1,831	1,029	101	28	510	9,000	-
Thailand	5,034	6,403	4,268	95	80	83,000	19,00
Vietnam	2,800	2,680	11,800	23	455	75,000	34,00
Australia	23,500	-	2,720	164,00	600	56,000	-
Fiji	160	-	30	-	60	2,000	-
New Zealand	7,999	-	428	64,800	1,000	9,000	-
Papua New							
Guinea	123	-	1,523	2	17	4,000	-
Samoa, W.	27	-	65	-	-	1,000	-
Tonga	8	-	65	-	11	-	-
Vanuatu	105	-	79	-	12	-	-
Asia-Pacific							
Total	405,114	135,654	408,860	471,697	273,250	3,663,000	466,00
World	1,265,976	140,648	889,141	1,171,403	521,974	9,661,000	528,00
Indian livestock							
Population as (%)							
Per cent of world	15.90	53,75	1.25	4.86	20.78	2.13	1.70

Animals are also classified on the basis of their nature of rumination (regurgitation). Some of them are listed below:

Non-ruminants (having single stomach)	**Ruminants (All ruminates; number of stomach may vary from usual of FOUR)**
1. Elephant	1. Cattle
Asian	Humped
African	Humpless
2. Rhinocerus	2. Buffalo
Indian	Asian
White	African
Black	Indian
3. Hippopotamus	3. Sheep
4. Horse	4. Goat
Domestic	
Wild	
Pony	
5. Zebra	5. Nilgai
Gravy	
Plains	
Mountain	
	6. Giraffe
	7. Deer
	8. Bison
	9. Llama (3 chambered stomach)
	10. Camels (3 chambered stomach)
6. Ass	Among species of the family *camelidae* which includes camels, vicunas, alpacas, llamas, the omasum is missing as it has been anatomically modified into small tubular and is structurally a part of the abomasums. Surprisingly enough, the reticulum contains glandular cells.
7. Pig	
8. Man	
9. Dos	
10. Monkey	

Taxonomy, Domestication and Animal Husbandry in India, China, Pakistan and Bangladesh

Taxonomy and Terminology

Taxonomy is the branch of biological science that attempts to classify living forms into some sort of order and system which can be used to relate "hierarchize" or otherwise identify them and establish their relationship to each other and to the totality of life. In general the taxonomic system is constructed according to the plan developed by Karl Von Linne in the middle part of 18th century:

1. Kingdom	2. Class	3.Family	7. Species
2. Phylum	4. Order	5. Genus	

For example, a human belongs to the Kingdom *Animal,* Phylum *Chordata,* Class *Mammalia,* Order *Primate,* Family *Hominidae,* Genus *Homo,* Species *sapiens.*

Zoological Classification (Taxonomy) of Common domestic animals

Kingdom	-	Animal
Phylum	-	Chordata
Class	-	Mammalia
Order	-	Primate
Family	-	Hominidae
Genus	-	*Homo*
Species	-	*Sapiens.*

Endocrine organs	Hormone	Source of hormone	Chemical nature	Tissue affected	Physiological function
(1)	(2)	(3)	(4)	(5)	(6)
Intermediate Lobe of Pituitary	Melanocyte stimulating hormone (MSH)	Basophilic-cells	Poly-peptide	Melanophore cells	1. Control of cutaneous pigmentation
Posterior Pituitary	Oxytocin	Supraoptic paravtricular nuclei of hypothalamus	Poly-peptide	Mammary gland and uterus	1. Release of milk 2. Contraction of uterus
	Vasoperssion or antidiuretic hormone	Supraoptic and para ventricular nuclei of hypothalamus	Poly-peptide	Peripheral blood vessels and kidney tubules	1. Contraction of blood vessels 2.Water reabsorption from kidney tubules
Parathyroid	Parathyroid hormone	Chief cells	Poly-peptide	Bone, kidney and intestine	1. Increases blood calcium level

contd...

Thyroid	Thyroxine	Thyroid follicles	Amino acid	All tissue	1. Increase rate of cellular metabolism
	Calcitonin		Poly-peptide	Bone and kidney	1. Decreases blood calcium level
Pancreas	Glucagon	Alpha cells of Islet of Langerhans	Poly-peptide	Liver	1. Raises blood sugar level
	Insulin	Beta cells of Islets of Langerhans	Poly-peptide	Muscles and adipose tissue	1. Lower blood sugar level
Ovary	Oestrogen	Ovarian follicles	Steroid	Mammary gland and genital tract	1. Development of female secondary characters 2. Maintenance of female secondary characters
	Progesterone	Corpus luteum	Steroid	Mammay gland and genital tract	1. Required for implantation of foetus
	Relaxin	Unknown	Poly-peptide	Cartilage and ligament of pelvic girdle	1. Relaxation of cartilage and ligament 2. Facilitate parturition

contd...

Testes	Androgens	Interstitial cells	Steroids	Accessory sex organs, testes	1. Development of secondary sex organ 2. Development of secondary sex
Kidney	Rennin and Angiotensin	Juxtraglomerular cells	Poly-peptide	Zonaglomerulosa	1. Increase blood pressure 2. Aldosterone Secretion
Adrenal Cortex	Glucocorticoids	Zona faciculata zona reticulate	Steroid	All tissues	1. Mobilization of body energy source 2. Increase blood sugar levels 3. Anti inflammatiory action
	Mineralo corticoids	Zonagromerulosa	Steroid	Kidney, all tissue	1. Salt balance 2. Water balance
Adrenal Medulla	Epinephrine	Chromaffin cells	Amino acids	Heart, skeletal muscles and liver	1.Increases heart strength 2. Increases blood sugar levels
	Nor Epinephrine	Chromaffin cells	Amino acids	Peripheral blood vessel and Smooth muscles	1. Constriction of peripheral blood vessel 2. Contraction of Smooth muscles and glands

Molybdenum	1. It is a component of xanthine oxidase and liver aldehyde oxidase 2. Xanthine oxidase catalyse the oxidation of many substrate including purines, aldehyde proteins and reduced diphos-phyropine nucleotide	The chief symptoms of molybdenum poising are: 1. Extreme diarrhoea and consequent weight loss. 2. Reduction in production. 3. Increase the copper deficiency	1. Cabbage, soyabean and peas. 2. Liver and glandular cell.
Luromium	1. Trivalent form chromium is needed for the normal glucose utilization. 2. It acts as a cofactor with insulin at the cellular level. 3. It reduces the level of serum cholesterol in the blood. 4. It involve in protein metabolism	Unknown	
Pluorine	1. Harden the tooth enamel and guard against tooth decaying. 2. Retards the osteoporosis in adults. 3. Enhance the growth in rats. 4. Plays an important role in the tissue formation of bone.	1. Excessive dental caries occur. 2. Higher level it is toxic to animals	
Shihhium	1. Acts as non-specific diet is toxic to anti-oxidant. 2. Protect against paroxidation in tissues and membrane. 3. Participate in the biosynthesis of ubi-quinone. 4. Participate in the hydrogen transport along with respiratory chain. 5. Check the degeneration and fibrosis of the pancreas in chicks. 6. It influences the absorption and retention of Vit. E and triglycerides.	Higher level of selenium in the animals.	

Important vitamins in animal nutrition

	Vitamins	Chemical name
I, Fat soluble vitamins		
	Vitamin A	Retinol
	Vitamin D_2	Ergocalciferol
	Vitamin D_3	Cholecalciferol
	Vitamin E	Tocopherol
	Vitamin K	Phylloquinone
II, Water soluble vitamins		
	Vitamin B_1	Thiamine
	Vitamin B_2	Riboflavin
	Niacin B_6	Nicotinamide
	Vitamin B_6	Pyridoxine
	Pantothenic acid	Pantothenic acid
	Biotin	Biotine
	Folacin	Folic acid
	Choline	Choline
	Vitamin B_{12}	Cyanocobalam
	Vitamin C	Ascorble acid

S.No.	Species of animals	Reccurance of heat after		Duration of heat	Range of gestation period (days)	Gestation period
		Non-Conception	Parturition			
1	Mare	21 days	7-11 days	6 days	305-400	340
2	Cow	21 days	30-60 days	6-30 hours	273-294	283
3	Buffalo	21 days	30-60 days	6-40 hours	300-309	304
4	Ewe	18 days	6-8 weeks	1-2 days	148-160	150
5	Goat	21 days	6-8 weeks	2-3 days	147-155	150
6	Sow	21 days	6-8 weeks	3-4 days	109-125	115
7	Bitch	Oestrus occurs twice each.		11-18 days	55-70	63
8	Cat	Year and they come into heat once during each season.		7-10 days	55-70	56

Normal body Temperature, Pulse rate and Respiration rate of Domestic Animals

Species of animals	Normal rectal temperature (F°)	Average rectal temperature (F°)	Normal pulse rate/minute	Normal respiration rate per minute
Horse	100.0-101.0	100.4	36-42	8-16
Foal	100.0-102.0	100.8	40-60	10-16
Ass	99.5-102.0	100.8	46-50	10-12
Mule	100.0-102.0	100.8	46-50	10-12
Calf	100.2-104.0	102.8	60-100	30-40
Cow	100.0-102.5	101.5	60-90	25-30
Ox	100.0-102.5	101.5	40-60	10-30
Buffalo	101.5-102.5	101.0	40-50	10-20
Sheep	102.0-104.0	103.0	70-90	20-30
Goat	102.0-104.0	103.0	70-90	20-30
Pig	102.0-104.0	102.6	60-90	10-20
Dog (Small)	100.0-102.0	101.8	90-130	16-30
Dog (Large)	101.0-102.0	101.0	70-100	16-30
Fowl	106.5-108.0	107.0	120-160	15-50
Cat	100.5-102.0	101.0	100-130	30-40
Camel	98.0-100.0	96.6	30-50	10-12
Elephant	96.6-98.0	97.6	25-35	0-10
Man	98.0-99.0	98.4	60-90	12-30

Reproductive Cycle of Domestic Animals

S. No.	Species	Age of pubery	Duration of ovulation	Time of ovulation	Period of maximum fertility	Gestation period	Sperm survival in uterus	Nature of heat cycle recurrance
1	Cow	2-2.5 days	8-24 hours Average 19 hours	14 hours after the end of oestrous	Last 8 hours of oestrous	282 days	20 hours	Polyoestrous Recur. 21 days
2	Buffalo	2.5 years 2.0 years-bull	8-36 hours average 32 hours	10-20 hours upto the end of oestrous	Last 8 hours of oestrous	307 days	Less than 24 hours	Polyoestrous Recurrance -21
3	Sheep	6-12 months	36 hours	Later part of oestrous	2nd half of oestrous	150 days	22-42 days	-
4	Goat	4-8 months	1-2 days	-	-	150 days	-	Second Polyoestraus Recur. 18-21 days
5	Pig	5 months	2-3 days	36 hrs. after onsent of oestrous and continued upto nesxt 12 hours	12 hrs. and 36 hrs. after the onset of oestrous	120 days	24 hours	Polyoestrous
6	Horse	2 years	3-14 days average 7 days	24 hours before the end of oestrous	Penultimate day of oestrous	336 days	2-3 days	Seasonally Polyoestrous (21 days)
7	Camel	3-4 years	1-7 days	-	-	406 days		Polyoestrous Recur. 10-20 days
8	Bitch	9-12 months	8 days	1-3 days after onset of oestrous	Next 3 days after the one set of oestrous	62 days	4 days	Monoestrous

Body Weight and Feed consumption of Broilers

Weeks	Body weight (g)		Feed consumption weekly (g)		Feed consumption cumulative (g)		Feed conversion ratio	
	Male	Female	Male	Female	Male	Female	Male	Female
1	107	105	82.5	81.5	82.5	81.5	-	-
2	250	230	182	161	264.5	242.5	-	-
3	460	410	329	280	593.5	522.5	-	-
4	700	600	434	357	1027.5	879.5	-	-
5	960	810	539	427	1566.5	1306.5	-	-
6	1300	1060	700	560	2266.5	1866.5	-	-
7	1670	1340	805	679	3071.5	2545.5	-	-
8	2060	1630	945	770	4016.5	3315.5	1 : 1.95	1: 2.03

Common Voices Produced and Places of Habitation

Species	Weaning Period	Puberty Age	Voices Produced	Place of Habitation
Horse	5-8 months	12-24 months	Neighing	Stable
Cattle	4-6 ″	12-18 ″	Bellowing	Byre
Buffalo	6-8 ″	12-18 ″	Bellowing	Byre
Sheep	3-4 ″	8-12 ″	Bleating	Pen (Bara)
Goat	6-8 weeks	8-12 ″	Bleating	Pen (Bara)
Pig	5-7 ″	4-5 ″	Grunting	Sty
Dog	5-7 ″	7-10 ″	Barking	Kennel
Cat	-	8-12 ″	Mewing	Kettery
Camel	-	3-4 years	Bleating/Bellowing	-
Elephant	-	20-25 ″	Trumpeting	-
Fowl	-	6 months	Crowing	Coop (Pen)

Common Terms used in different species of Livestock

Species	Weaning Period	Puberty Age	Voices Produced	Place of Habitation
Horse	5-8 months	12-24 months	Neighing	Stable
Cattle	4-6 "	12-18 "	Bellowing	Byre
Buffalo	6-8 "	12-18 "	Bellowing	Byre
Sheep	3-4 "	8-12 "	Bleating	Pen (Bara)
Goat	6-8 weeks	8-12 "	Bleating	Pen (Bara)
Pig	5-7 "	4-5 "	Grunting	Sty
Dog	5-7 "	7-10 "	Barking	Kennel
Cat	-	8-12 "	Mewing	Kettery
Camel	-	3-4 years	Bleating/Bellowing	-
Elephant	-	20-25 "	Trumpeting	-
Fowl	-	6 months	Crowing	Coop (Pen)

□□□

Chapter-3

Organic Manure and Biofertiliser

Organic manures have been used to improve and sustain soil productivity. These organic manures are classified into following 3-groups depending upon their nature and quality.

Organic manures

1.1 Bulky organic manures

1.2 Concentrated organic manures &

1.3 Biofertilizers

1.1 Bulky Organic Manures

- FYM (N - P - K) - 0.5-0.2-0.5
- Compost from farm litter (0.5-0.15-0.5)
- Compost from town refuse urban (1.4, 1.0, 1.4)
- VM (3.0-1.0-1.5)
- Night soil (5.5-4.0-2.0)
- Sludges (3-6-2.0-1.0)
- Sewage irrigation (1.5-3.5/0.78 to 4.0/0.3 to 0.6
- Green Manure

a) Legume	b) Non-Legume
i) *Crotalaria junecea* - 84.0 kg N/ha/season	i) *Cannabis sativa*

ii) *Sesbania aculeata* - 77.1 kg N/ha/season

iii) *Phaseolus tribulus* - 55.6 kg N/ha/season

iv) *Vigna sinensis* – 56.3 kg N/ha/season

v) *Cyamopsis tetragonoloba* – 62.3 kg N/ha/season

vi) *Melilotus alba* – 134.4 kg N/ha/season

vii) *Lathyrus sativus* – 61.4 kg N/ha/season

viii) *Trifolium alexandrinum* – 60.7 kg N/ha/season

ii) *Vernonia cineria* – Kojozira

iii) *Sorghum bicolor*

iv) *Zea mays*

v) *Helianthus annuus*

1.2 Concentrated Organic Manure

1.2.1 *Plant origin* (all oilcakes)

1.2.1.1 Non-edible oilcakes

- Castor cake (4.3-1.8-1.3)
- Cottonseed cake (3.9-1.8-1.6)
- Decorticated (6.4-2.8-2.9)
- Karanj (3.9-0.9-1.2)
- Mahua/ippi cake (2.5-0.8-1.8)
- Neem cake (5.2-1.0-1.4)
- Safflower cake

1.2.1.2 Edible oilcakes

- Coconut – (3.0-1.9-1.8)
- Cotton (decorticated) (6.4-2.9-2.2)
- Groundnut (7.3-1.5-1.3)
- Linseed (4.9-1.4-1.3)
- Niger (4.7-1.8-1.3)

- Rapeseed (5.2-1.8-1.2)
- Safflower (7.9-2.2-1.9)
- Sesame (6.2-2.0-1.2)

1.2.2 Animal origin

- Direct blood (10.0-1.0-06)
- Fish manure (4.0-2.0-0.3)
- Bird's guano (7.0-11.0-2.0)
- Poultry manure (3.03-2.63-1.40)
- Hoof and horn meal (14.0-)
- Raw bone meal (3.0-20.0)
- Steamed bone meal (1.0-25.0)

1.3. Biofertilizer

1.3.1 Soil Flora

Classification of Microorganisms

1.3.1.1 Bacteria

- Autotrophic Bacteria
  - a) Nitrifying Bacteria
    - *Nitrosomonas* or *Nitrosococcus*
    - *Nitrobacter*
  - b) Sulphur Bacteria - *Thiobacillus spp.*
  - c) Iron Bacteria - *Thiobacillus ferroxidans*
  - d) Manganese
  - e) Hydrogen - *Hydrogenomonas spp*
  - f) Co Bacteria
  - g) Methane bacteria - *Methanobacillus spp*
- Heterotrophic Bacteria

1. N. Fixing

a) Symbiotic eg. – *Rhizobium spp.*

***Rhizobius* spp.**

Sppnaure	Group Name	Crops
Rhizobium meliloti	Alfalfa group	Lucerne, Sweet clover
R. trifoli	Clover group	White clover
R. leguminosarum	Pea group	Garden, fieldpea, sweetpea, Bean, lentil
R. phaseoli	*Phaseolus* group	Garden & Kidney bean
R. japonicun	Soybean group	Soybean
R. lupini	Lupin group	Annual/Perennial lupins
R. (unnamed spp.)	Cowpea group	Cowpea, Limabean, groundnut (Peanut)

b) Non-symbiotic

i) Aerobic eg. - *Azotobacter chroococcum*
- *A. vinelandi*

ii) Anaerobic eg.
- *Clostridium paesteurianum*

2. Non-N-Fixing

i) Aerobic eg. *Ammonifiers* $(+O_2)$
-*Pseudomonas spp*
- *Bacillus spp*
- *Micrococcus spp*
- *Paracoccus spp*
- *Proteus spp*

ii) Anaerobic eg. *Denitrifiers* $(-O_2)$
- *Pseudomonas*
- *Bacillus spp*
- *Micrococcus spp*
- *Paracoccus spp*
- *Proteus spp*

iii) Cellulose decomposer
- *Cytophaga*
- *Sporocytophaga*
- *Angiococcus*
- *Polyangium*

1.3.1.2 Fungi

i) Moulds	ii) Yeasts	iii) Mushrooms
Cellulose decomposer	*Trichoderma, Aspergillus*	
Protein decomposer	*Fusarium, Mucor*	
Lignin decomposer	*Bacidiomycetes*	
Humus former	*Alternaria, Aspergillus*	

1.3.1.3 Actinomycetes

- Streptomycin
- Potato scab disease etc.

1.3.1.4 Algae

- BGA (Blue green algae) - *Cyanophyta, Anabaera, Nostoc, Cylindrospermum*
- GGA (Grass green algae) - *Chlorophyta*
- YGA (Yellow golden algae) - *Xanthophyta*
- GBA (Golden brown algae) - *Bacillariophyta (Diatoms)*

1.3.2. Soil Fauna

Macro

- Earthworm
- Moles
- Ants, centipedes, white ants
- Others millipedes

  Snails, mites, termites

Micro

Protozoa

- *Parameceium*
- *Euglena*
- *Amoeba*
- Nematodes

Organic Manure and Related Material

These are a very important class of nutrient sources for crops and a key component of integrated crop nutrition. All organics are not alike and differences between them are as large as those between fertilisers and manures.

Green Manures: These are green plants, mostly N-fixing legumes which are buried in the soil before sowing the main crop to add organic matter and nutrients present in the green manures. Such manures can be grown right in the main field (example *Dhaincha*, Sunnhemp, short-duration legumes) or green leaves from plants grown elsewhere can be brought in and buried (example Subabul). It is thus not always essential to grow green manure right where it is needed. A good 60-day crop of *Dhainclua* can add 60-80 Kg N plus other nutrients. Nutrient wise and from easy of decomposition, it is the best of organic manures. The problem is most farmers these days do not practice green manuring due to land, labour and cost involved and also due to relatively cheaper urea fertiliser.

Farmyard Manure (FYM): This is the common name for well rotten animal dung mixed with urine, shed litter, left over fodder etc. It is available wherever animals are kept but amount available for manure is much less than the dung produced because most of the dung is used for fuel in the rural houses. If it is used in biogas plant. a farmer can have both fuel and manure. FYM is very widely recommended but mostly farmers are seen to apply it for growing vegetables, potato, sugarcane. maize etc. Its composition can vary by 100% from lot to lot.

Composts: These are obtained from systematically decomposed organic material through several processes. Composts produced in the villages are called rural composts and those from urban/city refuse, garbage or wastes are called urban composts. These are richer than FYM in nutrient content but still come under bulky organic manures. These can be enriched to a certain extent by fortification with fertilisers and minerals such as urea, rock phosphate etc. Vermicompost is produced with the help of earthworms who have the ability to digest and discharge a variety of organic wastes.

Crop Residues: These include products such as straw, stover, sticks etc. after the grain is, removed, or leaf trash of sugarcane, hair (coir) of coconuts, peels of fruits and vegetables, shells of groundnut, husk of coconut and rice etc. Essentially any plant parts remaining after the main produce is taken away is a residue. Crop residues are particularly rich source of potash because most of the potash absorbed by the crop is left behind and as soil mulches for conserving soil moisture and lowering soil temperature.

Oil Cakes: These are a special category of organic manures and are residues after the oil has been extracted from seeds. These are more concentrated than most plant-based organic manures because these are derived from nutrient-rich seeds and not nutrient depleted straw. Due to their high price, cakes of edible oils are used mostly for cattle feed and export while cakes of non-edible oil (castor, neem, etc.) can be used for manure. *Neem* and *Karanj* seed cake have a special role in improving the efficiency of urea fertiliser. The same amount of urea produces higher yield when it is mixed/blended with powdered neem cake.

Manures Derived from Animal Body: These include' products derived from the blood, hoof, flesh etc. of dead animals and are generally called animal meals. Other products include bone meal (both raw and steamed).

Biogas slurry, Sewage and Sludge: The biogas slurry is the thick liquid discharged by biogas (gobar) plants. It has whatever was present in the dung used to make biogas and is a good manure. From convenience point of view it can be spread to fields near the biogas plant. Sewage and sludge are obtained from human wastes. Sewage is the mixture of these wastes with water as it arrives in sewage treatment plants. It has lot of water and can be used for irrigation after different degrees of treatment according to the crop for which it is to be used. Too heavy irrigation with these waters can lead to soil sickness and does not serve the purpose. Sludge is the semi-solid material left at the bottom of treatment tanks. Before using these, it is essential to find out their composition and correct application (loading) rates. Some suggestions given by Dr. Jawarkar of NEERI, Nagpur are given below:

Waste water	Suggested crops in order of preference	
Primary treated	Cash crops	Cotton, jute, sugarcane, tobacco
	Essential crops	*Citronella, Mentha,* lemon grass
	Cereals and pulses	Wheat, rice, greengram, blackgram, sorghum, pearlmillet
	Oilseeds	Linseed, sesame, castor, sunflower, soybean and groundnut
	Fruit crops	Coconut, banana, citrus, sapota, guava, grapes, papaya, mango
	Vegetables	Brinjals, beans, lady finger etc. These should be exclusively cooked before eating
Secondary treated		All crops listed above
		All crops including vegetables borne near the soil surface, but consumed after cooking only
Secondary treated and disinfected		All crops without restrictions

Biofertiliser

These are not traditional fertilisers in that they do contain plant nutrients. Most biofertilisers are bacteria which fix atmospheric N into ammoniacal form so that it can he used by plants. Such biofertilisers are *Rhizobium, Azotobacter, Azospirillum.* and *Cyanobacteria* (BGA). *Azolla* is a fern in the cavities of which the *Cyanobacteria* fix N. Two other important biofertilisers are the phosphate-solubilising microorganisms and the mycorrhizae which help the plant roots to gather more nutrients from a bigger soil volume.

Rhizobium: Among the biofertilisers the most important is the *Rhizobium*. It works specifically (symbiotically) with leguminous plants. It is thus an important contributor of N to the nutrition of pulses, legume forages, oil crops such as groundnut and soybean. Blue-green algae and *Azolla* can serve as biofertiliser for wetland/f1ooded rice. Their role is not much in upland rice or at unfavourable temperature regimes. Mycorrhizae is (least used at the practical level, partly due to the difficulty of multiplying it on a large scale).

Any strain of *Rhizobium* can not be used to inoculate any legume seed. These are specific well-worked out matches and only the appropriate strain (culture.), should be used. Normally the biofertiliser packet indicates the crop for which that product is meant. For example, the *Rhizobium* species which develop symbiotic (mutually-beneficial) relationship with particular legumes are:

Crop	Matching *Rhizobium*
Soybean	*Rhizobium japonicum*
Groundnut, horsegram	*Rhizobium sp.* (cowpea miscellary)
Greengram, cowpea, pigeonpea	
Chickpea	*Rhizobium leguminosarum*
Bean	*Rhizobium phaseoli*
Lentil, Pea	*Rhizobium leguminosarum*
Lucerne	*Rhizobium meliloti*
Berseem	*Rhizobium trifolii*

Blue Green Algae (BGA): The actual number of BGA species is very large but a very few of these are actually N-fixing. Among the important N-fixers are most species of the *Anabaena, Nostoc* and *Cylindrospermum* genera. These ideally function at temperatures upto 38°C and can fix 20-30 kg N/ha for the paddy

Azospirillum: It has 4 species of which *A. lipoferum* and *A. brasilense* are the most important. These are recommended (not everywhere) for a large number of upland crops and can fix 15-25 kg N/ha in a season).

Azotobacter: Like *Azospirilum*, it does not need to live with a plant (like *Rhizobium* does with legumes) in order to fix N. That is why these are called free-living. Out of 7 species of *Azotobacter, A. chroococcum, A. vinelandii* and *A. beijerinkia* are more important. Its use as recommended in general as in case of *Azospirillum* and N contribution is also similar in general.

Azolla: It has seven species of which the most common is *A. pinnata*. It fixes N symbiotically with *Anabaena azollae* (BGA). It works best in submerged rice fields within 38°C and ideally under adequate supply of phosphorus in the soil medium. *Azolla* can be grown both as a traditional green manure and alongwith rice plants in the main field.

Phosphorus Solubilising Biofertilisers: These can be bacteria or fungi. Among bacteria, the important ones are *Bacillus megatherium, B. polymyxa, Pseudomonas striata, P. liquifaciens* etc. Among fungi, these are *Aspergillus awamori, A. fumigatus, Penicilium digitatum, P. lilacinum* etc.

The Mycorrhizae belong to the genera such as, *Glomus, Gigaspora, Acaulospora, Scleocystis* and *Endogone.* They do not fix nutrients but enhance the ability of roots to gather more of them.

The technical aspects of biofertilisers are discussed in greater detail in the following FDCO publications:

1. Biofertiliser Technology, Marketing and Usage
2. Fertilisers, Organic Manures. Recyclable Wastes and Biofertilisers.

The above discussion briefly presents an account of plant nutrients and their carriers. These differ greatly in their physical and chemical properties, bulk, local availability, material and application costs. It thus very important to know the advantages and disadvantages of each at the field level and then develop the most suitable integrated packages.

Classification of Organic Manures

Organic manures have been used to improve and sustain soil productivity. These manures are classified into –

i) Bulky organic manures.
ii) Concentrated organic manures.
iii) Bio-fertilizers.

(i) Bulk Organic Manures

These are farmyard manure (FYM), Compost, Vermimanure (VM) and green manure (GM) added in a large quantity because of their bulk, mainly to improve physical condition of the soil, replenish and keep its humus status, maintain the optimum conditions for the activities of soil microorganisms and make good a small part of the plant nutrients removed by the crops or otherwise lost through leaching and soil erosion. They thus supply practically all the elements of soil fertility, that the crop require, though not in adequate amounts and in right proportions. The plant food elements contained in manures are released in an available form upon decomposition by soil microorganisms. The main role of these manures is to add organic matter and humus content. The properties and the roles of organic matter and humus in soil are as follows :

- Coarse organic matter on the surface reduces the impact of the falling rain drops and thus reduces surface runoff and erosion.

- Easily decomposable organic residues lead to synthesis of complex organic substances that bind soil particles into structural units called aggregates, which help to maintain a favourable condition of aeration and permeability (Pedology).
- Organic matter increases water-holding capacity of the soil.
- Organic matter adds and serves as a reservoir of essential nutrients which are released in harmony with the needs of the plants.
- It produces organic acids (H_2CO_3), that help in dissolving available K, P, micronutrients etc. forms in soil.

In a summary the effect of bulky organic manures on soils is three fold

i) Bulky organic manures contain nutrients in small quantities, therefore large quantities of them need to be applied per ha. Besides the major nutrients, bulky organic manures also contain traces of micro/minor/trace nutrients.

ii) Bulky organic manures increase organic matter content and hence improve the physical properties of soils. This effect is very important in case of most of our arable land such manures increase the humus content of soils at least temporarily, and consequently the H_2O holding capacity of soils is increased and the drainage of clayey soils is improved, $CO_2+H_2O \rightarrow H_2CO_3$ (carbonic acid) $\rightarrow$ Acidity.

iii) Bulky organic manures provide food for soil microorganisms. This increase activity of microbes which in turn help to convert unavailable plant nutrients into available forms.

Farmyard Manure (FYM)

The term farmyard refers to the decomposed mixture of dung and urine of farm animals alongwith the litter (bedding material) and left over material from roughages or fodder fed to the cattle. Farmyard manure collected daily from the cattle shed consists mainly dung and part of the urine soaked in the refuse. Newly collected and stored FYM is fresh as against well decomposed FYM which has been stored for a sufficient period of time to allow its decomposition to completion. On an average, well-rotted farmyard manure contains 0.5 per cent N, 0.2 per cent P_2O_5 and 0.5 per cent K_2O.

Trench method of manure preparation

Trenches of 6m long, 2m wide and 1m deep are dug. The material consisting of dung and urine soaked in it, is collected daily and placed in these trenches. Additions are normally applied in sections of one metre length. When each section is filled upto a height of about 0.5m above the ground level. The top of the heap is rounded off to the shape of a dome and plastered with shiny of cow dung and earth. Before plastering it is important to apply 4-5 buckets of H_2O in the pit. Plastering conserves moisture and nitrogen and also prevents housefly nuisance. The manure pit should be protected from sun and rain. The manure becomes ready for use in about 4-5 months after plastering. Normally 10-25 cartloads of manure are applied per ha. A cartload of manure measuring 9 cubic metres weighs about 0.5 tonnes. It is possible to obtain 5-6 tonnes of good quality manure per year per head of cattle.

Precautions

- The manure trench/pit should be far away from the residential house to avoid the fly population and diseases.
- The cowshed floor should be cemented with concrete.
- Out grazing of animal should be avoided.
- The plastering should be done completely.
- The trench should be sheltered by shed.
- The trench or pit should be cemented.
- Bioactivator/preservator should be added.
- Caking of cowdung should be avoided.
- The well rotted farmyard manure should be applied to the field and mixed with soil immediately.
- Volatilization losses of N in the form of NH_3 should be avoided by not heaping the manure into field before sowing of the crop.
- The raw FYM should not be applied to field otherwise weed and termite problems damage the crop.
- The gobar gas slurry should be applied to the kitchen garden specially for vegetable crops.
- The dung, urine and litter should be removed from shed daily.

- The litter used for animal should be a good quality for soaking the urine and comfortable to animal.
- Handling of manuring should be proper.
- The cattle in the shed should be kept in a proper manner.

FYM can be studied into following heads :

1. Chemical composition of fresh excreta of animals
2. Quality of excreta produced per year by different types of animals
3. Factors influencing the composition of manure
   i) Source of manure.
   ii) Food of the animals.
   iii) Age and condition of the animal.
   iv) Function of the animal.
   v) Manner of storage.
   vi) Nature of litter.
4. Losses during handling and storage of farmyard manure
   A) Losses during handling.
      i) Loss of liquid portion or urine.
      ii) Loss of solid portion or dung.
   B) Losses during storage.
      i) By leaching.
      ii) By volatilization.
5. Improved methods of handling farmyard manure
   i) Trench method of preparing farmyard manure.
   ii) Use of gobar gas - compost plant.
   iii) Proper field management of farmyard manure.
   iv) Use of chemical preservatives.
6. Supply of plant nutrients through farmyard manure

7. Crop which respond to farmyard manure
   - i) Crops which show profitable response to application of farmyard manure.
   - ii) Crops less responsive to application of FYM.
8. Potential supplies of farmyard manure and the quantities at present
9. It serves as a source of energy for the growth of microorganisms as well as food for such life as earthworms, ants and rodents
10. Organic mulches reduce evaporation losses
11. Optimum temperature can also be maintained by organic mulches

Principal green manure crops suitable for growing *in situ*

Both Legumes and non-legumes are used as green manure crops in India. Legumes due to their additional advantages of adding atmospheric nitrogen to the soil, are mostly used as green manure crops, while non-legumes are used to a limited extent. List of common leguminous green-manuring crops are :

Botanical Name	Common Name	Growing Season
Crotalaria juncea	Sunnhemp	*Kharif*
Sesbania aculeata	Dhaincha	*Kharif*
Phaseolus trilobulus	Pillipesara	*Kharif*
Vigna radiata	Mung	*Kharif*
Vigna sinensis	Cowpea	*Kharif*
Cyamopsis tetragonoloba	Guar	*Kharif*
Melilotus alba	Senji	*Rabi*
Lathyrus sativus	Khesari	*Rabi*
Trifolium alexandrinum	Berseem	*Rabi*

The most commonly used non-leguminous GM crops are :

Cannabis sativa, Vernonia cinerea, Sorghum biolor, Zea mays, Helianthus annuus.

Green manure crops

i) *Sunnhemp* - This is the most outstanding green manure crop. It is well suited to almost all parts of the country, provided the field is not swampy and the area receives sufficient rainfall or has an assured irrigation. It is extensively used with sugarcane, potatoes garden peas;

ii) *Dhaincha* - It occupies the second place next to sunnhemp for green manuring. It has the advantage of growing under adverse conditions of drought, waterlogging, salinity and acidity.

iii) *Sesbania rostrata* - It is aquatic leguminous crop which have nodules both on the stem and roots. It was introduced in India in the1980's from the IRRI, Philippines. It is a tropical legume which thrives well under flooded and waterlogged conditions, producing aerial nodules on the stem. Due to its profuse stem nodulation, it gives ten times more nodules than most of the legumes.

Technique of green manuring (*in situ*) - The maximum benefit from green manuring can not be obtained without knowing - (i) when the green manure crops should be grown, (ii) when they should be buried in the soil and (iii) how much time should be given between the burying of a green manure crop and the sowing of the next crop.

Time of sowing green manure crops (*in situ*) - The normal practice usually adopted is to begin sowings immediately after the first monsoon rains.

The stage of green manure crop at burying - It can be generalized that a green manure crop may be turned in at the flowering stage. The maturity of the green manure crops take about seven to eight weeks from the time of sowing to attain the flowering stage.

Green leaf manuring - *Glyricidia maculata, Sesbania speciosa, Tephrosia purpurea, Indigofera tesmania (shrub). Trees - Pongamia pinnata, Terminalia arjuna, Cassia tora.*

Regions not suitable for GM - Dryland farming (arid and semi-arid) receiving < 62.5 cm of annual rainfall, on very fertile soil in good physical condition. In areas where *rabi* crops raised on conserved soil moisture.

Green manuring

Green manuring can be defined as a practice of ploughing or turning into the soil undecomposed green plant tissues for the purpose of improving physical structure as well as fertility of the soil. There are two types of green manuring -

i) **Green manuring *in situ*** - In this practice, green manure crops are grown and buried in the same field which is to be green manured either as a pure crop or as intercrop with the main crop. The most common green manure crops grown under this system are - sunnhemp, dhaincha pillipesara (*Phaseolus trilobulus*) and guar (*Cyamopsis tetragonoloba*) etc. (Northern India).

ii) **Green-leaf manuring** - Green-leaf manuring refers to turning into the soil green leaves and tender green twigs collected from shrubs and trees grown on bunds, wastelands and nearby forest areas. The common shrubs and trees used are - *Glyricidia* (*Glyricidia maculata*), *Sesbania speciosa,* Karanj (*Pongamia pinnata*) etc. (Eastern and central India).

Advantages of green manuring

- It adds the organic matter to the soil. This stimulates the activity of soil micro-organisms.
- The green manure crops return to the upper top soil, plant nutrient taken up by the crop from deeper layers.
- It improves the physical condition (structure) of the soil.
- It facilitates the penetration of rain water, thus decreasing run off and erosion.
- The green manure crops hold plant nutrients that would otherwise be lost by leaching.
- When leguminous plants, like-sunnhemp and dhaincha are used as green manure crops, they add N to the soil for the succeeding crop.
- It increases the availability of certain plant nutrients, like phosphorus (P_2O_5), Ca^{+2}, K^+, Mg^{2+} and $Fe^{+2/+3}$.

Brown manuring

Brown manuring is a technique to grow *Sesbania* in standing rice (inter crop) crop and then on its flowering (25-30 DAS) stage destroyed with the help of herbicide (2, 4-D @ 1.5 kg ha^{-1}) for manuring. After killing the colour of *Sesbania* residue become brown so it called brown manuring. Brown manuring practice introduced where sesbania crop can survive easily and @ 20 kg seed ha^{-1} is broadcasted three days after rice transplanting and allow to grow for 30 days and dried by spraying 2, 4-D ethyle easter which supplied upto 35 kg ha^{-1} N, dry matter. It also control broad leaf weeds, higher yield by 4-5 qha^{-1} due to addition of organic matter in low fertile soils. There is potential of up-scaling of rice + *Sesbania* irrigated system including salanity affected canal irrigated area.

Disadvantages of GM crops

i) Under rainfed conditions, it is feared that proper decomposition of the green manure crop and satisfactory germination of the succeeding crop may not take place if sufficient rainfall is not received after burying the green manure crop.

ii) Since green manuring for wheat means loss of *kharif* crop, the practice of green manuring may not be always economical. This applies to regions where irrigation facilities are available for raising *kharif* alongwith easy availability of fertilizers.

iii) The cost of growing green manure crops may be more than the cost of commercial nitrogenous fertilizers.

iv) An increase of diseases, insects and nematodes is possible.

v) A risk is involved in obtaining a satisfactory stand and growth of the green manure crops, if sufficient rainfall is not available.

Desirable characteristics of GM crop (*In situ*)

- It should be a legume with good nodular growth habit indicative of rapid nitrogen fixation under even unfavourable soil conditions.
- It should have little water requirements for its own growth and should be capable of making a good stand on poor and exhausted soils.
- It should have a deep root system which can open the subsoil and top lower regions for plant nutrients.
- The plant should be of a leafy habit capable of producing heavy tender growth early in its life cycle.

- It should contain large qualities of non-fibrous tissues of rapid decomposability containing fair percentage of moisture and nitrogen.

Input Requirements

Under Rice - Tablepea - Onion, cropping sequence.

Problem (1) A research scholar of Agronomy wants to give an organic treatment in his research trial. He has to carry out his experiment under RBD with three replication and ten Treatment - T_1, T_2, T_3, T_4, T_5, T_6, T_7, T_8, T_9 and T_{10}

T_1 - 100 % RND as FYM.
T_2 - 125 % RND as FYM. } Farmyard manure
T_3 - 150 % RND as FYM.

T_4 - 100 % RND as VM.
T_5 - 125 % RND as VM. } Vermi manure
T_6 - 150 % RND as VM.

T_7 - 100 % RND as PM.
T_8 - 125 % RND as PM. } Poultry manure
T_9 - 150 % RND as PM.

T_{10} - 100 % through urea.

Calculate the needed amount of Nitrogen for the cropping sequence Rice - Tablepea - onion if the plot size is 4×2 m^2 and % N of FYM = 0.5%, VM = 2% and PM = 3%. (Recommended N for rice - 120, tablepea - 20 & Onion - 50 kg/ha^{-1}).

A. Needed Amount of FYM

$T_1 \rightarrow 0.5 - 100$ ($\therefore$ when 0.5 % N)

120 - 100/0.5 × 120 = 24,000 kg ha^{-1} or 19.2 kg N plot^{-1}.

(T_2 & T_3 in same manner)

B. T_4 - 1.0–100/1 × 20 = 1200 kg ha^{-1}, 1200/1000 × 10.4 = 12.186 plot^{-1}.

So on 100/1 × 120 = 1200 kg ha^{-1}, 1200/1000 × 10.4 = 12.186 plot^{-1}.

[∴ when 1% N in VM]

$$T_4 = \frac{100}{1} \times 120 \times 8 = 96{,}000 \text{ kg N ha}^{-1}$$

Plot Size = 8 m^2, then = $\frac{96000}{10000} \times 8 = 76.8 kg$

C. T_7 = 3 - 100/3 × 4 = 4000/10000 × 10.4 = 3.48 kg plot.

Rest-so on 100/1×8 = 1200 kg ha^{-1} 1200/1000×10.4 = 12.186 kg N plot^{-1}.

D. T_{10} =100/46×120 =>60.86 kg ha^{-1} or 260.86/10000×10.4=0.271 kg plot for three Replication it will be = 0.271×3 = 0.8138 kg N plot^{-1}.

Chemical composition of different organic Manures

	OM	N%	P_2O_5	K_2O %
1.	FYM	0.5	0.2	0.5
2.	VM	3.0	1.0	1.5
3.	PM	3.03	2.63	1.40

2. Calculated the needed amount of Farmyard Manure in respect to N, P, K in groundnut crop of a ha.

Chemical composition of FYM = N 0.5%, P_2O_5 0.2, K_2O 0.5

(1) 0.5 - 100 FYM

1 - 100/0.5

120 - 1000 × 120 = 5000 Kg ha^{-1}.

How much P_2O_5/K_2O has been supplied - ?

P_2O_5 = 10 kg ha, 1000 - 2/1 × 24

48 kg ha^{-1} P_2O_5 and 120 kg ha^{-1} Nitrogen or 120 kg K_2O_5 ha^{-1}.

or ratio - 4 : 10 : 10 5 : 2 :5 or 5: 2 :5

5 + 2 + 5 = 12

12 - N - 5 × 2000 = 10000 kg ha^{-1}.

12 - 2 × 2000 = 2000 kg ha^{-1}.

12 - 5/12 × 24000 = 10000 kg ha^{-1}.

Or

0.5 - 100/0.5 × 120 = 24000 kg ha^{-1}.

0.2 - 100/0.2 × 12

0.5 -

Problem (2) Calculate the needed amount of nutrients from the following particulers groundnut cultivation -

1. Recommended dose of NPK - 25 - 35 - 70 kg ha^{-1}.
2. Chemical composition of FYM = 0.5% 0.2% 0.5%
3. Other chemical fertilizers are - SSP and Kcl.
4. Ca $(Mo_3)_2$ - 15% N.

Calculate different organic or inorganic nutrients or fertilizer or manure for groundnut crop -

1. 0.5 - 100

   1 - 100/0.5

   25 - 100×10/5 × 5 = 5000 kg FYM or

   Supplied Nitrogen = 5 × 5 × 1/1 = 25 kg N ha^{-1} N.
2. 0.2 P_2O_5 - 100/0.2

   1 - 100/0.2

   35 - 100/2 × 5 × 35 = 17500 kg FYM P_2O_5
3. 0.5 - 100/0.5

   or - 100/5 × 2 ×70 = 14000 kg ha^{-1}.

□□□

Chapter-4

Vermiculture and Vermicomposting

Classification of Red Wiggler Worm

Kingdom – Animalia

Phylum - Annelida

Super phylum – Lophotrochozoa

Superclass : Clitellata

Class : Chaetopoda

Subclass : Oligochaeta

Order : Opisthopora/Haplotaxida

Suborder : Lumbricina

Superfamily : Lumbricoideae

Family : Lumbricidae

Scientific Name: *Eisenia foetida* (red wiggler worm)

Genus : Eisenia

Species : Foetida

1. Polychaeta (Polychaetes or bristle worm) – *Arenicola* (*lugworm*), *Neris* (*ragworm*)
2. Oligochaeta (Oligochaetes or Earthworm)
3. *Hirudinea* (leeches)

Etymology : From the Latin *Annellus* a little ring.

Characteristics of Annelida

1. Bilaterally symmetrical and vermiform.
2. Body has more than two cell layers, tissues and organs.
3. Body cavity is a true coelom, often divided by internal septa.
4. Body possesses a through gut with mouth and anus.
5. Body possesses 3 separate sections, a prosomium, a trunk and a pygidium.
6. Has a nervous system with an anterior nerve ring, ganglia and a ventral nerve chord.
7. Has a true closed circulatory system.
8. Has no true respiratory organs.
9. Reproduction normally sexual and gonochoristic or hermaphoditic.
10. Feed a wide range of material.
11. Live in most environments.

Phylum Annelida

Earthworms

Fig. 4.1

Earthworms are enormously important in the construction and fertility maintenance of the soil and were described by Aristotle as 'the intestines of the earth'. Numerous soil scientists have been equally fascinated by the amount of work done by them, Charles Darwin said of them "It may be doubted whether there many other animals which have played so important a part in the history of the world as these lovely organised creatures".

Earthworms are miniature top soil factories, earthworms make soil, all other living things eventually pass through an earthworm on the way to becoming soil, and it is likely that nearly every atom in your body has been in an earthworm's stomach before it was part of you, (some salts and ions, and some water are the exceptions). Earthworm castings are rich in all the minerals necessary for plant growth in a water soluble form so that they are immediately available for plant use. All the soil you have ever seen, at least in temperate climes, has passed through the stomachs of numerous earthworms to become what it is. The best way to deal with those unwanted earthworm casts on your lawn is to collect them and use them as potting compost, no manufacturer can make anything better suited to the growth of seedlings.

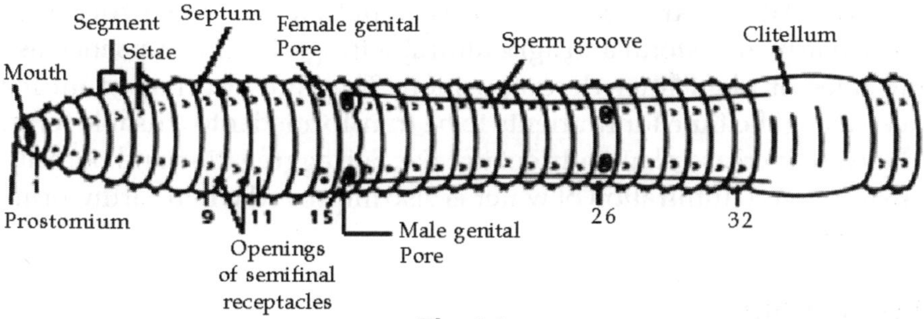

Fig. 4.2

It has been scientifically proven that earthworms increase the productivity of many soils, in some cases doubling or tripling crop yields. They do this by improving the structure of the soil, by bring nutrients up to the surface layers of the soil from deeper down and assisting in the break down of organic matter in, or on the surface of the soil.

This has resulted in a huge market in worms springing up in the second half of the 20th century in Western Europe and USA. In many cases this has been good, but in others, ignorance has resulted in the efforts and money spent being wasted. Soil have a definite worm carrying capacity which relates directly to the amount of organic matter in, and regularly added to, them (yearly in natural environment). Adding more worms than the soil can carry, or adding worms to soils poor or deficient in organic matter will result in only a small temporary increase in fertility resulting from the nutrients released from the dead worms and not from the work they have done. Further more not all worms are suitable for all soils and not all worms are active in soil creation, in particular *Eisenia foetida,* a commonly sold species which is very useful in composts and dung recycling is useless in, and will not survive in, ordinary field soils.

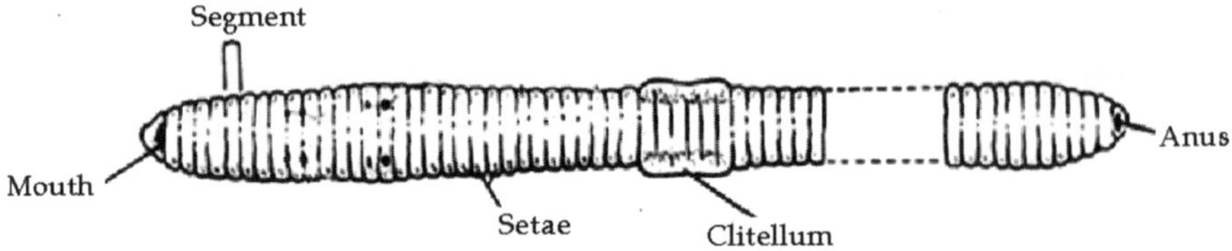

Fig. 4.3

Benefit(s): Are often seen as beneficial to soil processes, particularly in pastoral and agricultural settings. Earthworms increase the rates of plant litter decomposition. They are also important for decreasing the time for nutrients to be transformed into a usable form. They also increase plant nutrient uptake. Soil aggregation and porosity are improved. Infiltration of water is also improved when earthworms are present.

Vermiculture

Vermiculture biotechnology invoking the use of earthworms as versatile natural bioveactors for effective recycling of non-toxic organic wastes to the soil, resulting in soil improvement and sustainable agriculture. They effectively harness the beneficial soil microflora, destroy soil pathogens and converts organic wastes into valuable products such as biofertilizers, vitamins, enzymes, antibiotics growth hormones and proteinous worm biomas. Hence, we can call the earthworms as 'Nature fertilizer factories'.

- Earthworm's gut is an effective tubular bioreactor, with raw materials (feed) entering from one end and the product (castings) coming out through the other end.
- They maintain a stable temperature through normal temperature regulation mechanisms, thus accelerating the rates of bioprocesses and preventing enzyme inactivation caused by high temperatures.
- Gizzard is a novel colloidal mill in which the feed is ground into particles smaller than 2 microns, giving thereby an enhanced surface area for microbial processing.
- They have an in house supply of enzymes such as protease, lipase, amylase, cellulose and chitinase, which bio-degrade complex triomolecules, into simple compound utilizable by the symbiotic gut microflora.

- Earthworms have a butt in O_2 plant which can separate aerial O_2 by chemical absorption into blood haemoglobin.
- Castings contain nutrients in a balanced proportion and are rich in vitamins, enzymes, antibiotics and growth hormones.

Vermiculture biotechnology can be fruitfully utilized to gain several benefits:

To Farmers

i) Less reliance on purchased inputs leading to low cast of production.

ii) Enhancement of soil productivity.

iii) Recycling of organic wastes is achieved.

To Industries

There is a cost effective pollution abatement technology.

To Environment

i) Wastes create no pollution, as they become valuable raw materials for the soil biotechnology processes.

ii) There is more ground H_2O recharge and less ground water depletion.

iii) Soil salinization is reduced with low soil erosion and runoff.

To National Economy

i) Lesser import of agrochemicals, saving valuable foreign exchange.

ii) Savings on subsidies.

iii) Boost to rural economy.

iv) More export of agricultural produces with lower pesticide residues.

v) Reduced wasteland formation.

Vermiculture process involved in vermicomposting Vermi-composting involves the steps like :

1. Selection of site and construction of tank.

2. Selection of suitable species.
3. Collection and half decomposition of substrates.
4. Preparation of vermi-bed and watering.
5. Release of earthworms.
6. Putting the different layers of substrates.
7. Covering with gunny bags/cloths.
8. Control of insect pests.
9. Turning upside down at one month interval.
10. Harvesting followed by separation of worms and cocoons.
11. Drying under shade.
12. Winnowing and separation of undecomposed materials and packing and marketing.

1. **Selection of Site and Construction of tank** - raised, under shade, there should a shed and which having the watering source, sun and rain avoided.

   Vermicomposting can be done in pots, bins, trays, barrels, temporary or permanent tanks, pits and heaps but in all the cases there should be a cover to protect the unit from sun and rain. Direct sunlight not only reduces the activities of the worms but increase gaseous loss of nitrogen (NH_3) while rainfall causes leaching loss of nutrients and growth promoting substances. Excess of H_2O due to rain can create anaerobic condition and kill the worms. However, there should be a drainage hole at the bottom of the unit and water should be applied after proper examination so that there is minimum drainage of water. Beginners can start with a temporary unit of 2m×1m×0.6m dimention which can be constructed simply by putting bricks in layers over a cemented floor or a sheet of polythene. The innermost side of the unit should also be covered with polythene sheets before tilling it with bedding materials or substrates. Such a unit can be dismantled with less cost at the time of harvesting and assembled again for the next cycle.

2. **Selection of suitable species** - The species of earthworm should have qualities like :

a) Preference for eating concentrated half decomposed animal dung and wide range of organic waste.

b) Utilizing only around 5% of the foods eaten for its own maintenance and excreting the rest as cast.

c) High consumption, digestion and assimilation rate and.

d) High multiplication rate.

e) Worm species should be tolerant to disease.

f) The culturing techniques should be simple enough to adopt.

g) Worm should be efficient convertor of plant or animal biomass to body proteins, so that its growth rates are high.

h) The worm should have wide adoptability (tolerance) to environmental factors (capability to live in varying temperature conditions).

i) The worm should produce large numbers of cocoons that should not have long hatching time, so that multiplication and organic matter conversion is fast.

j) Growth rate, maturity from young one to adult stage should be fast.

k) Worm should have compatibility or tolerance with other worms as would add to productivity of bio-mass (worms) and conversion rate at different strata (layers) of organic matter, i.e. faster composting.

l) Worms on introduction in substrate, should have least inactivity period (vermistabilization period).

m) The worms should feed near the surface of organic matter.

n) They should have ability to bear transportation shock at their cocoon stage (resting phase).

Important species for vermicomposting

Eisenia foetida, Lumbricus rubellus, L. terrestris, Eudrillus spp., *Lampito mauriti, Octochaetona striata, Dravida willsi, D. surrensis* and *D. thurstoni*

3. Collection and half decomposition of substrates and vermicomposting materials

The biologically degradable and decomposable organic waste are used in vermiculture and vermicomposting. Commonly used composting materials are as follows :

All types of biologically decomposable non-poisonous wastes like animal dungs, agricultural wastes, orchard leaf litter, processed food wastes, waste paper, cotton cloths, city refuge and biogas slurry can be used for vercomposting. However the pH should be around neutrality or some time or egg cells may be added and use of acidic substances, pet excreta, excess of cruciferous leaves should be avoided. Dung or urine and poultry litters being acidic in nature should be used after considerable lapse of time. The substrates should be mixed with 20-50% dung around 10% of leguminous materials and rest of cereal and other straws. RP may be added @2.5% P_2O_5 to increase the phosphorus content in the vermicompost. However, use of paddy straw has been found to increase only the K_2O content. Use of greasy, coloured and glazy papers should be avoided. Rice husk being full of Silica takes more than 6 months to decompose and its residue in the final product reduces consumer acceptance. It is also to be kept in mind that the materials fed, to the animal from whom the dung is collected and the area where the plants required for substrates were grown, should not have contained more of heavy metals mentioned earlier.

Collection of agricultural waste should follow the practices like -

a) breaking of large lumps.

b) cutting of bigger plant parts into smaller ones.

c) exposing to sun to reduce excess moisture content.

d) application of 4% aqueous solution of neem pesticides to kill insects if any and

e) treatment with lime dust to increase pH.

Half decomposition of the materials is done by heaping the above mixture with sufficient moisture content for 21 days during which its temperature increases to 65°F and falls down to ambient temperature at the end.

4. **Preparation of vermibed and watering** - Vermibed is prepared by putting pebbles, sand and loamy soil one above the other having 2 inches of thickness each, at the bottom of the unit. Alternatively coir or any plant refuge which does not decompose easily can also be used. This should be watered followed by putting well decomposed FYM of 4 inches thickness over it.

5. **Release of earthworms** - Earthworms are to be released over the top of animal dung. For the above mentioned size of 2m × 1m × 0.6m 2kg worms (which may be 2000 to 2500 depending upon the size of the worms) are to be released. However the number of worms could be doubled when they are available in more quantities and compost is necessary to be prepared within a month instead of the normal time of two months.

6. **Putting the different layers of substrates** - Above the layer of FYM the half decomposed materials are to be kept to fill up the pit. When there are more number of pits half decomposition can be done in some of there in which different materials can be put in layers till the pit is filled up and worms are to be released after the thermophilic stage of half decomposition is over.

7. **Covering with gunny cloths** - After filling up of the unit it is to be covered preferably with an old wet gunny cloth to reduce loss of moisture and encourage the activities of the worms at the surface. Even if the worms are epigeic in nature they may not like to come to the top if the upper portion is exposed to light. When worms come to the upper surface they eat away the weed seeds present there which is very much essential for maintaining weed free condition in the final product. If the unit is not covered, one can find some weed seeds to germinate; while other remain viable - an undesirable character for a good quality vermicompost. Intermittently the unit is to be watered to maintain moisture content of 40 to 50%.

8. **Control of insect pests** - Care should be taken to control ants and other insects by applying turmeric powder alone or mixing with neem cake, over and around the unit. Dried leaves of *Lantana camara* could be used around the tank as an alternative, chemicals should not be used as that would make the product unfit to be used under organic farming.

9. **Turning upside down at one month internal** - It is very much necessary to maintain aerobic condition inside the heap for production of quality vermicompost. It helps in growth of beneficial microorganisms. Therefore, the heap should be thoroughly mixed at an interval of 30 days. However, when more of dung is used, mixing becomes necessary at an internal of 15 days.

10. **Harvesting followed by separation of worms and cocoons** - Harvesting of vermicompost is done when the compost looks dark brown and soft. The vermicompost is piled under sun for 3 to 4 hours so that the verms go down and forma ball which can be separated for further use. But in small units when the top portion of the unit is fund to be covered with few inches of casts these could be collected for selling at a premium price.

11. **Drying under shade** - After separation of the worm ball as narrated earlier, the compost should be dried under shade for reducing its moisture content to around 30%. This helps to preserve all the nutrients and growth factors present in the compost.

12. **Winnowing and separation of undecomposed materials** - The compost is passed through a sieve of 2-3 mm diameter so that the cocoons and the undecompposed materials could be separated for further use.

13. **Packing** - Packing can be done in beautifully labelled polythene for attracting the consumers. Its date of packing, amount of nutrients present, the dose of application and the date before which it should be used for getting the maximum benefit should also be written for specific type of consumers. Chemical analysis for every batch may be done separately which should be done in recognized laboratories. Home scale gardens require as small as 1kg packets and may require packets of 5kg or larger.

14. **Use of vermicompost** - Farmers prefer to use vermicompost for the garden crops or crops having an aesthetic or high value when the quantity is limited. In flowering pots it is applied at the rate of 200g/pot having a single plant. This dose is applicable once in three months.

□□□

Chapter-5

Biological Properties and Soil Health

Classification of Microorganisms

Bacteria	Fungi	Actinomycetes	Algae
I. Autotrophic Bacteria	1. Parasitic	Streptomycin	Green
A. Nitrifying Bacteria	2. Saprophytic	Potato scab	Blue green
B. Sulphur Bacteria	3. Symbiotic	disease etc	Yellow Green
C. Iron Bacteria		*Mycorrhiza*	Diatoms
D. Manganese Bacteria		a. Ectotrophic	
E. Hydrogen Bacteria		b. Endotrophic	
F. CO Bacteria			*Phoma*
G. Methane Bacteria			*Pythium*

II. Heterotophic Bacteria

A. N-fixing Bacteria

1. Symbiotic e.g. *Rhizobium*
2. Non-Symbiotic
   a. Aerobic e.g. *Azotobacter*
   b. Anaerobic e.g. *Clostridium*

B. Non N-fixing Bact.

1. Aerobic e.g. Ammonifiers
2. Anaerobic e.g. Denitrifiers

Autotrophic Bacteria

Obtain carbon from the CO_2 of atmosphere and energy from the oxidation of simple carbon compounds or from inorganic substances. These organisms are self-sustaining i.e. they can built complex compounds required for their living processes by using very simple inorganic substances. Specific groups of autotrophs are capable of oxidizing ammonia, nitrate, sulphur, iron, manganese, hydrogen, CO, or CH_4 (Methane). The most important groups of autotrophs are those that oxidize ammonia to nitrite to nitrate known as nitrifying bacteria.

Heterotrophic Bacteria

1. Comprise of majority of soil bacteria.
2. Consume organic substances for their energy and carbon source.
3. Primarily decompose cellulose and other carbohydrates, proteins, fats and waxes.
4. Functionally they bring about mineralization of organic matter through hydrolysis and oxidation and release nitrogen, phosphorus and other nutrients in form of available to plants.
5. N-fixing bacteria are another example of heterotrophs. These can work under aerobic conditions or under anaerobic conditions.

Fungi

Fungi may be parasitic, saprophytic or symbiotic. Parasitic fungi produce plant diseases like cotton root, rot, many kinds of wilts, rusts, blights and smuts.

Saprophytic fungi obtain their energy from the decomposition of organic matter.

Symbiotic fungi live on the roots of certain plants and both fungus and plant are mutually benefited.

Fungus are specially useful in soils as they break down the somewhat resistant cellulose, lignins and gums, as well as more readily decomposed sugars, starches and proteins. A large part of the slowly decomposing soil humus consists of the dead remains of fungal hyphae.

Mycorrhiza:- Meaning fungus root.

A symbiotic association of fungal mycelium and roots of certain trees and shrubs. It is presumed that mycorrhiza aid the host plant in the absorption of certain nutrients.

Forest nurseries that have been established to raise tree seedlings that are not native to the area usually need an artificial inoculation of the suitable mycorrhiza the reasoning behind this practice is that in a new region the compatible mycorrhiza are usually not present in the soil. There are two types of mycorrhizae based upon their manner of growth.

A. **Ectotrophic** - Formed by members of *Agaricales* including mostly mushroom type fruiting fungi. They grow as thread like filaments into small roots between the root cells, but not into the cell. They increase the absorbing surface of the tree roots, ectotrophic *mycorrhiza* is exhibited by pines, spruces, oak, elms, beech, hickories, chestnut and birches.

B. **Endotrophic** - Formed by species of *Phoma* and *Pythium*. The fungal hyphae penetrate the root cells upon dying mycorrhizal tissues are absorbed and utilized by the growing trees. e.g. sweet gums, poplars, maples, laurels, azaleas, rhododendrons and orchids.

Actinomycetes

1. Taonomically and morphologically related to both fungi and bacteria, but have recently been classified as bacteria.
2. Characterized by branched mycelia similar to fungi and when the mycelia break into fragments they resemble to bacteria.
3. Common examples are- Streptomycin, Aureomycin, Tetramycin and Neomycin.
4. They thrive best when there is ample fresh organic matter, when the soil is neutral to slightly acidic and when soil moisture is fairly abundant. However when the soil is fairly dry the grow better than fungi.
5. Their primary function is in decomposing organic matter, especially cellulose and other resistant forms.
6. Potato scab diseases an actenomyces can be readily controlled by keeping the pH of the soil below 5.0

Algae

Soil algae are microscopic chlorophyll bearing organisms. The main groups are :

1. Green 2. Blue green 3. Yellow green 4. Diatoms

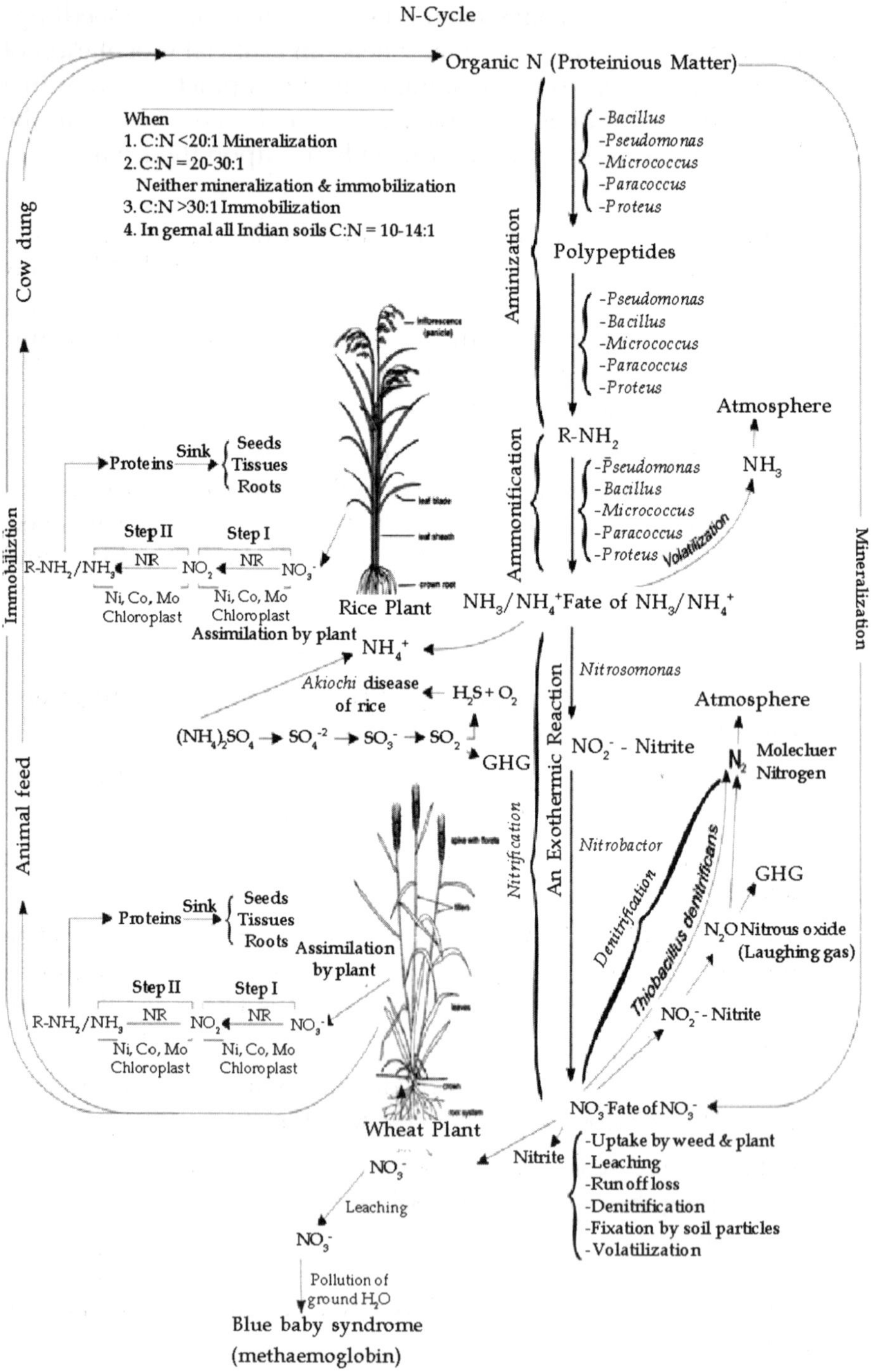

Fig. 5.1. Microbial Transformations

Algae grow best in most fertile soils. The green colour of the soil surface following the application of commercial fertilizers is due to an increase in their number.

The affect of algae on plant growth is :

a. Add organic matter to the soil as they are self sustaining organisms.

b. Improve soil aeration - especially in rice fields by excretion oxygen.

c. Fix atmospheric nitrogen -Blue green algae is capable of fixing atmospheric nitrogen. The best pH range for working of blue green algae is 7 to 8.5. In rice fields they contribute significant amount of N from the atmosphere. Even in desert soil the blue green algae are and high N content of many surface soils.

Nitrogen

The transformation of nitrogen in the soil can be grouped into two :

1. Gains

A. Mineralisation

1. Ammonification

2. Nitrification

B. Atmospheric Nitrogen fixation

1. Fertilizers

2. Rain water

3. Microbial

2. Losses

A. Immobilization

B. Denitrification

C. Crop removal, weed removal

1. Ammonification

The transformation of organic nitrogenous compounds into ammonia is called ammonification. Heterotrophic organisms which form the bulk of soil population break down organic matter into simple substances available for plant nutrition or for further oxidation by Autotrophic organisms.

The process involes a gradual simplification of compounds. The process is of two types - A. Aerobic ; B. Anaerobic.

In aerobic process heterotrophs consume oxygen and release CO_2. The steps are :

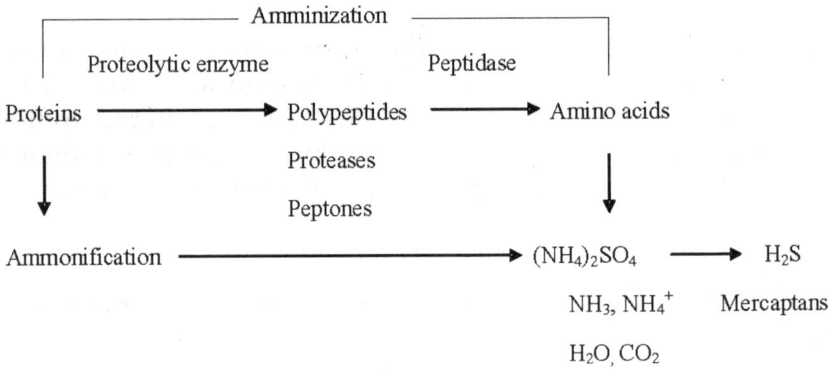

Anaerobic process (Putrefaction)

Proteolytic enzyme

Proteins → Polypeptides → NH_3, Amines, CO_2, Organic acids, Indole.

Organisms

Aerobic bacteria, fungi, Actinomycetes and certain facultative and strict anaerobes.

For highly proteinaceous material - species of *Pseudomonas, Bacillus, Clostridium, Serratia, Micrococcus* readily decompose pure proteins. Under neutral to alkaline environment- Bacteria under acid conditions proteins are broken down chiefly by fungi.

Ammonification occurs as a result of action of enzymes product by microorganisms. The action is chiefly hydrolytic and oxidative. For best result well drained aerated soils are best suited for rapid ammonification.

Part of the ammonia or amino acids so product are utilized by the microorganisms for their own body building proteins. They are organic compounds as their N - source. But if the organic matter as a very wide C : N ratio then the microorganisms utilize the NO_3 or NH_3. If it is available in the soil (Immobilization). Thus the immediate effect of addition of quantities of easily oxidisable carbohydrates like molasses or straw would be immobilization of NO_3 or NH_3, if no nitrogen from external source is not added with such materials. The ideal C : N ratio of accomplish mineralization is 10:1.

Nitrification

The oxidative sequence involved in the conversion of ammonia to nitrate is termed nitrification.

Nitrification is an aerobic process carried out by Autotrophic bacteria. It occurs in two steps :

$$\text{Organic nitrogen} \rightarrow NH_3 \xrightarrow[\text{Energy}]{\text{Nitrosomonas}} NO_2 \xrightarrow[\text{Nitrification}]{\text{Nitrobacter}} NO_3 + \text{energy}$$

Ammonification

The detailed steps and mechanism are :

Nitrosomonas

$$NH_3 \xrightarrow{\frac{1}{2}O_2} NH_2OH \xrightarrow{-2H} \tfrac{1}{2}\,HON = NOH \xrightarrow{\frac{1}{2}O_2} HNO_2 + 79 \text{ cal.}$$

Hyponitrite Hydroxyl amine

(Its presence is not definitely established).

Nitrobacter

$$HNO_2 \xrightarrow{H_2O} HON(OH)_2 \xrightarrow{-2H} HNO_3 + 43 \text{ cal.}$$

Conversion of NO_2 to NO_3 is faster as compared to that of NH_3 to NO_2, accumulation of NO_2 is toxic to plants.

The source of carbon of these organisms is CO_2 of atmosphere or HCO_3^-.

Ion in soil

Nitrosomonas oxidize 35-70 moles of NH_3 and *Nitrobacter* 70–100 moles of NO_2 per mole of carbon assimilated.

The overall reactions can be summarized as :

$2\ NH_3 + 3O_2 \rightarrow 2\ HNO_2 + 2H_2O + 79$ cal. - *Nitrosomonas*

$2\ HNO_2 + O_2 \rightarrow 2NHO_3 + 43$ cal. - *Nitrobacter.*

Factors affecting nitrification

1. Soluble organic matter - Depresses nitrification but the presence of heterotrophic organisms keeps the soluble organic matter at a low level. Thus heterotrophs and autotroph works for their mutual benefit.
2. Soil reaction - Slow in acid and alkali soils but favoured by neutral pH.
3. Phosphorus supply - Phosphorus addition favours the process. The efficiency of super-compost (compost + SSP) is explained by this.
4. Oxygen supply - The nitrifying organisms are aerobic and have a high requirement of O_2, therefore favoured by well drained, aerated soils.
5. Soil texture - Nitrification occurs at the colloidal surface where abundant NH_4^+ is absorbed and where bacteria adhere. Favored by loamy or clayloam soils and slow in sandy soil.
6. Moisture - 50% WHC suitable for the reaction.
7. Temperature - Greatest at 37°C, stops below 5°C and above 55°C.

Nitorgen Fixation

N-fixation is one of the three most important processes, the other two are photosynthesis and respiration. Biological N-fixation in one of the chief sources of nitrogen for soils. Plant growth leaching and activities of denitrifying organisms all tend to remove the soil nitrogen reserve though plant residues returned to soil for decomposition add to the soil status. Still considerable quantity of soil N is build up by biological N-fixation.

There are two main classes of organisms that fix atmospheric nitrogen :

A. Non symbiotic

B. Symbiotic

A. Non-symbiotic N-fixing organisms

The free living organisms capable of utilizing molecular nitrogen include. *Azotobacter, Clostridium* as the main genera of bacterial population besides a wide range of heterotophic bacteria, autotrophic organisms, blue green algae are capable of fixing atmospheric nitrogen nitrogen non symbiotically.

Azotobacter

It is a most important N-fixing organisms responsible for non symbiotic nitrogen fixation. It is a strongly aerobic and oxidizes carbohydrates and other such sources of carbon completely to CO_2 to H_2O The efficiency of *Azotobacter* to fix atom N does not exceed 15-20 mg N/g of carbohydrate oxidized. *Beijerinkia* spp. Are more tolerant of acidic soils.

Condition (Habitat)

Azotobacter is favoured by a neutral reaction, ample phosphorus, aerobic conditions and moist well drained soils. *Azotobacter* is susceptible to $PO_4^=$ deficiency. The properties is utilized in detecting the P-deficiency of soils in *Azotobacter* plaque tech. Other nutritional requirement of *Azatobacter* are :

1. Vitamin-B
2. Nucleic acids
3. Ca
4. Mo
5. Mn-(Mn compound accelerate N-fixation by *Azatobacter*).

Clostridium

Clostridium pasteurianum etc. works in anaerobic conditions. This organism is less efficient than *Azatobacter*. Fixation capacity of nitrogen by this organisms is 2-3 mg/g of carbohydrate fermented. The low efficiency of *Clostridium* can be attributed to the fact that while anaerobic fermentation releases only a small amount of energy, aerobic oxidation produced tremendously large amount of energy which help to fix more atmospheric N.

Non-symbiotic N-fixation occurs on the roots of corn, wheat, millet, sorghum, rice and species of earthen higher plants.

Mechanism of Non Symbiotic Nitrogen Fixation

The partial pressure of nitrogen at which N fixation take place non symbiotically at half the maximum velocity is :

For	*Azatobacter*	-	0.02 atm.
	Rhodospirrillum	-	0.02 atm.
	Nostoc	-	0.02 atm.
	Clostridium	-	0.03 atm.
	Redclovers	-	0.05 atm.

Higher the concentration of N in the soil higher will be fixation rate. Practically no fixation takes place below 0.02 atm. and above 0.05 atm. of N.

$$N_2 + 6H \xrightarrow{\text{Nitrogenase}} 2\,NH_3$$ The sequence is

$$N = N \xrightarrow{\text{Nitrogenase (Enz.)}} N = NH \xrightarrow{\text{Enz.}} NH - NH_2 \xrightarrow{\text{(Enz.)}} 2NH_3$$

OR

$$N = N \xrightarrow{\text{Enz.}} \underset{\text{Diamide}}{NH = NH} \xrightarrow{\text{Enz.}} NH_2 - NH_2 \xrightarrow{\text{Enz + H2}} 2NH_3$$

The enzyme nitrogenase contains Mo and hence the need of Mo for non-symbiotic N-fixation.

Symbiotic Nitrogen Fixation

The classical example is the symbiosis between bacteria of genus *Rhizobium* and plant of leguminceae. The association is sited in root of the host plant in the form of nodules. Most legumes differ from other plants in that they use, in addition to the soluble N to the soil, the atmospheric N, made available to them by the bacteria living in the nodules of their roots.

Nodule : These are rounded outgrowths on the roots and root hairs of the legume plants.

Mature nodule consists of a central core containing non dividing cells of the bacterial tissues (which is the site of N-Fixation) surrounded by the host tissue having double the usual chromosome number. This tissue of the host plants is called disomatic.

The nodules are of two types

1. Effective
2. Non-effective

1. **Effective Nodules :** The effective nodules are capable of a successful host guest symbiosis and result into atmospheric nitrogen fixation. The effective nodules are pink, large sized and less bunched then the ineffective nodules. The pink colour is due to the presence of leghaemoglobin. The concentration of leghaemoglobins directly linked with amount of N-fixed by the bacteria. This also suggests the need of iron and oxygen for larger proportion of N-fixation. The leghaemoglobin plays a molar role in the nitrogen metabolism of the *Rhizobium* species in the plant.

2. **Non effective nodules :** More branched, brown coloured no leghaemoglobin in them. Incapable of fixing N_2.

Usually a certain strains of *Rhizobia* species can form nodules on limited groups host plants. These group host plant is called the cross inoculation group.

Process of nodule formation

The first step involves release into root zone of the plant excretory product like tryptophan, which are stimulatory to the bacteria. The *Rhizobia* then aggregate at distinct sites adjacent to roots. The *Rhizobia* product B indole acetic acid. B-IAA which causes root hair deformation.

The infection starts at the root hairs. The *Rhizobium* synthesizes and excretes one or more product provoking deformation of legume root hair. These compounds include a nucleic acid, polysaccharide and/or proteins. The root hair wall at this site of synthesis of the above compounds gets irritated, softened and invaginates. This invagination then continues to develop into tube like structure called infection thread. This tube contains the *Rhizobia* bacteria. Finally the nodule and the bacteria are ultimately released into symbiont's cytoplasm to multiply further. Immediately following the release of these bacteria in the host cell a period of rapid cell division of the host cells surrounding the infection are take place giving rise to a swollen round structure of nodule. The final structure is a central core, containing the *Rhizobia* and surrounding cortical containing plant vascular system.

The host cells of the central portion of the legume contains double the chromosome number characteristic of the host.

The central core containing the *Rhizobia* contains them in the form of non-dividing resting cells of *Rhizobium* called bacteroids. This central core is the site of fixation of atmospheric nitrogen. Bacteroids are non-motile, non-viable derivative cells original infecting *Rhizobia*. Bacteroids have a complete nitrogen system capable of fixing N_2. In addition to the nitrogenase complex, bacteroid have pigment called leghaemoglobin which reversibly combines with oxygen in a manner similar to that of animal haemoglobin. The oxygenated leghaemoglobin then transports the oxygen under low free oxygen tension to the sites of the bacteroid oxidative phosphorylation to be used for generation of ATP.

Fixation of N by Non legumes

The symbiotic fixation of N_2 is not confined to the plants of the legume family only eight genera of the angiosperms comprising some 190 spp of widely distributed trees and shrubs also produce nodules. These plants do fix significant quantities of atmospheric nitrogen and this fixation is confined to the nodules. The responsible organisms have not been isolated.

Mechanism of symbiotic N-fixation

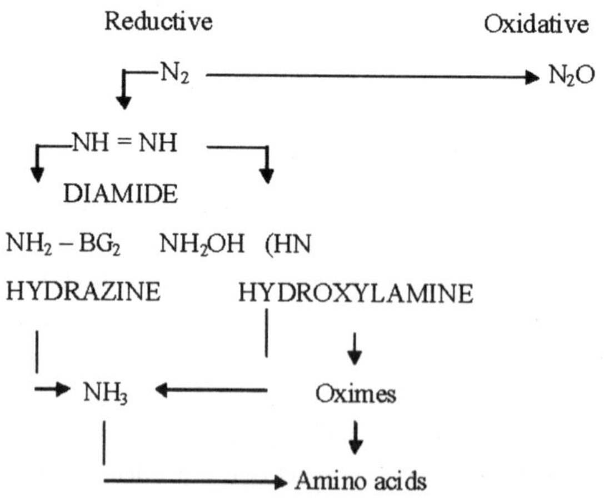

The fixation of inert N_2 is accomplished by the enzyme nitrogenase. This enzyme lowers the activation energy (The energy needed to cause the reaction to occur). The fixation process for both symbiotic and non-symbiotic nitrogen fixation are almost similar. The first step is adsorption of molecular nitrogen (N_2) by the enzyme nitrogenase.

N=N bond is broken to form NH=NH and NH_2-NH_2 to form $2NH_3$.

Fate of ammonia

All the above take place the nitrogen is bonded to the enzyme. The NH_3 combines with some decomposition products of carbohydrates, organic acids to form amino acids which ultimately condense to form peptide linkages and them proteins. One such example is formation of glutamine from á-keto glutaric acid. The reaction involves enzymes like glutamic dehydrogenase, glutamine synthetase and one more is carbamyl kinase to produce carbonyl phosphate. The reactions are:

Glutamic dehydrogenase

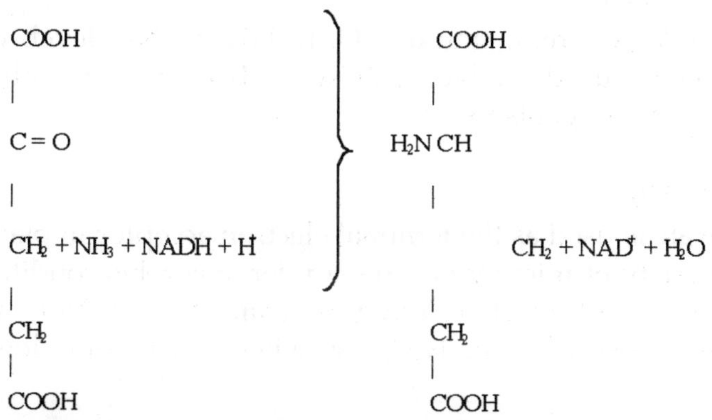

α- Keto glutaric acid L-Glutamic acid

Glutamine synthetase

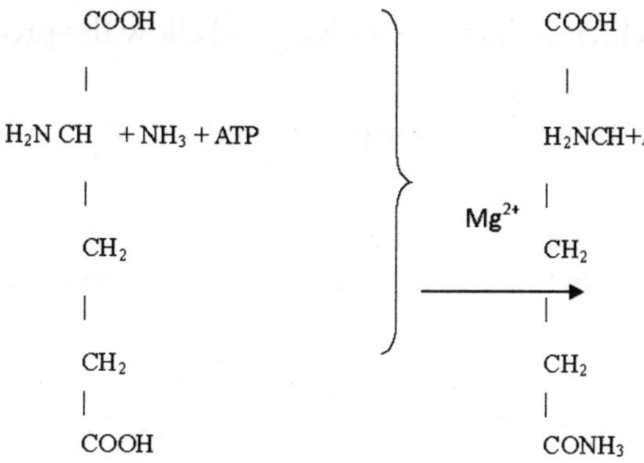

Glutamic acid Glutamine

Carbamyl kinase

$$NH_3 + CO_2 + ATP \overset{Mg^{2+}}{\leftrightarrow} H_2N\text{-}\underset{\underset{O}{||}}{C}\text{-}OPO_3H_2 + ADP$$

Carbamyl phosphate

Denitrification

It is microbial process whereby NO_3^- is reduced to gaseous products. It is one of the major factors of losses of nitrogen from soil.

The process under which nitrates are utilized by microorganisms fall into two general classes:- 1

(a) Nitrate Assimilation

Denotes biological reduction of NO_3 to NH_3 or NH_2 level, with the products being used for biosynthesis of N-containing cellular constituents of microorganisms.

(b) NO_3-Respiration

In this nitrate is used as the terminal electron acceptor in place of oxygen by a variety of microorganisms, under anaerobic conditions. These organisms obtain their oxygen from NO_3 or NO_2 with accompanying release of N_2 or N_2O, e.g. when soil become logged water.

Organisms involved

Two distinct group of micro-organisms have been noted :

1. Those organisms which reduce NO_3 to NO_2 and allow the product to accumulate.
2. These are true denitrifiers and completely reduce NO_3 to N_2 or NO or N_2O.

The capacity for denitrification is limited to bacterial population only and so far no fungi or actinomycetes have been reported to carry out this loss.

Since common examples are : sp. of *Pseudomonas, Micrococcus* (sp.- *Denitrificans*) *chromobacter, Bacillus* etc. *Autotrophs* like *Thiobacillus denitrificans, Thiobacillus thioparus* etc.

Mechanism

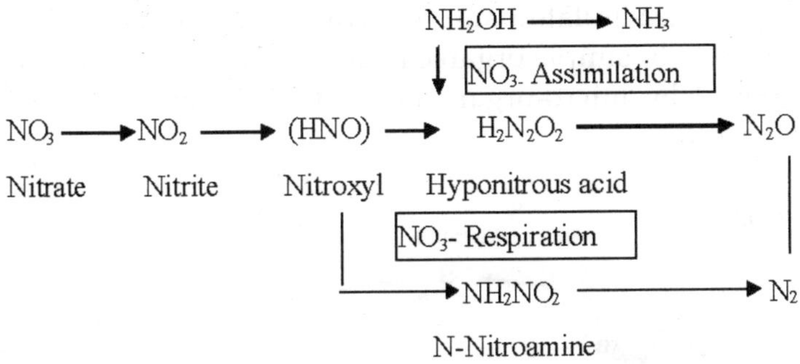

Factors

1. Denitrification occurs only when the supply of oxygen is restricted for the bacteria thus in well drained aerated soil no denitrification will take place but if the moisture level is above 60% of water holding capacity the rate of denitrification increases.

2. Adequate supply of organic substrate is required as the energy source for bacteria.

3. Presence or absence of growing crops as they compete for NO_3 with each other. Losses of N through denitrification can be minimized by keeping adequate drainage and pH level 5.5 to 7.0.

Transformation of Sulphur

The total sulphur content of tropical soil is generally lower than that of temperate soil because of lower organic matter and greater leaching. Sulphur occurs in soil are :

Inorganic forms - Sulphate of Mg, Ca and Na, Sulphide (S), Sulphite (SO_3), H_2S.

Organic forms - S - amino acids like cystine, cysteine and methionine, the others are peptide glutathione, glucosides, alkaloids, thiamine, biotin, thioctic acid, thiocynate, mercaptans, taurine, thiourea, ethereal, sulphate, thiosulphates, thiosulphides, organic sulphates etc.

Sulphur compounds undergo many transformation as a result of activities of plants animal and microorganisms, and to a lesser extent non biological actions.

Plants assimilate sulphur as $SO_4^=$ form soil and smaller amount at S-amino acids and direct assimilation of atmospheric SO_2. Animals utilize this plant assimilated sulphur converting reduced sulphur compound to SO_4^- or simple organic forms. Plant and animal remains are decomposed by micro-organisms in the soil with liberation of inorganic sulphur.

Sulphur Cycle

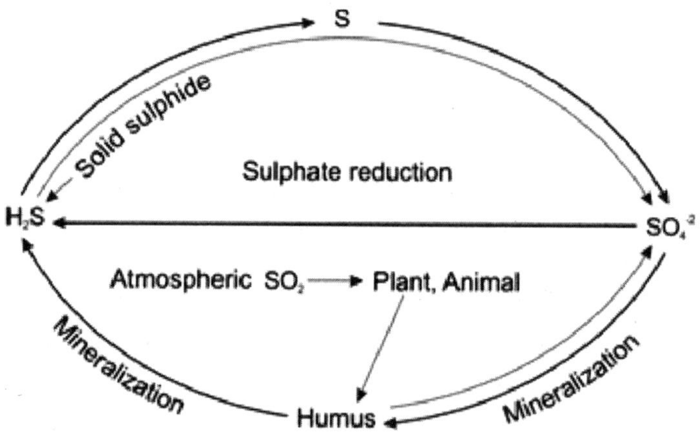

The major microbial process are :

1. Mineralisation : Break down of organic sulphur molecules to inorganic sulphate.
2. Immobilisation : Conversion of inorganic sulphur to organic sulphur compound.
3. Oxidation : Conversion of reduced inorganic sulphur compounds to sulphate.
4. Reduction : The dissimilatory reduction of sulphate and intermediate compounds to sulphide (S).

Mineralisation

Plant and animals residues containing sulphur compound act substrate for microbial activity. Prominent among these are S-containing proteins, amino acids - cystine, cysteine, and methionine. The peptide glutathione, glucosides, alkaloids, thiamine, biotin, thioctic acid, thiocynater, mercaptans, taurine, ethereal sulphate, sulphates, thiosulphates and sulphides etc.

Mechanism

Soil microorganism produce a variety of proteolytic enzymes which can hydrolyse peptide and ester bond. The proteinaceous compound are eventually broken down to amino acids.

Proteins —Proteolytic enz.→ Polypeptides —Peptidase→ Peptides amino acid

sulphate esters in plants, animals and microorganisms

sulphohydrolases

Arylsulphatases

(Hydrolysis)

Under aerobic condition to end products are SO_4^{-2} and elemental sulphur.

Under anaerobic condition end products are mercaptans, H_2S.

Immobilisation

Such organic matter which is low in sulphur but is energy rich is added to soil then the soil microorganism make use of the inorganic S for their own body building. Thus the inorganic sulphur is converted into organic forms. To avoid it sulphur fertilization is necessary. (Like that of N)

Mechanism

The process is assimilatory sulphur reduction involving enzymatic reaction of

SO_4^{-2} SO_3^{-} to S^{-} compound and elemental sulphur.

The high energy low sulphur organic matter can immobilize added sulphur from 38.06 to 45% as it has been recovered from the falvic acid fraction of the humus.

Organisms Involved In Mineralisation and Immobilisation

Mostly heterotrophic organisms, strict and facultative anaerobes, and all those organisms involved in the mineralization and immobilization of N are capable doing so for sulphur also.

Factors

Soil properties, temperature, moisture, S content of organic matter, rhizospheric activities, plant roots etc.

Sulphur Oxidation

Formation of sulphate and other sulphur compound from more reduced form sulphur both organic and inorganic in soil is sulphur oxidation.

S-oxidation is carried out by 4 groups of organisms.

1. Chemoautotrophic bacteria of genus *Thiobacillus*.
2. Poorly defined group of heterotrophic bacteria, fungi and actinomycetes.
3. Colour has filamentous bacteria of genera *Baggiatoa, Thiotrix* and *Thioplaca*.
4. Photosynthetic sulphur bacteria belonging to families Thiorhodaceae and Chlorobacteriaceae.

Only the first two groups are found in soils. *Thiobacillia* obtain all the energy required for their growth from the oxidation of inorganic sulphur compound to sulphates. All of the organisms of the genus are obligate autotrophs with exception of *T. novellus* which is facultative autotroph capable of developing on organic substrates.

T-denitrificans is capable of utilizing NO_3 anaeorobically in place of oxygen.

T-ferroxidans oxidize Fe^{++} iron in addition to sulphur for its energy.

Other differences between members of thiobacilli include the degree of acid tolerance and range of organic sulphur compound which they can oxidiise.

Most of the Thiobacilli are capable of oxidizing sulphide, elemental sulphur, thiosulphate, tetrathiovate and in some cases sulphide to sulphate.

Some typical reaction are :

Thiobacillus thioxidans

$$S + \frac{1}{2} O_2 + H_2O \rightarrow 2\ NaHSO_4$$

Sod. Thiosulphate

T. thioparus

$$Na_2S_4O_6 + Na_2CO_3 + \frac{1}{2}O_2 \rightarrow Na_2SO_4 + CO_2$$

Sodium tetrathionate

T. denitrificans (anaeorobically)

$$5S + 6\ KNO_3 + H_2O \rightarrow 2K_2SO_4 + 2\ H_2SO_4 + 3N_2$$

Other species of *Thiobacillus* carry out this oxidation as :

$$H_2S + 2O_2 \rightarrow H_2SO_4$$

$$2H_2S + 2O_2 \rightarrow 2S + 2H_2O + 2H_2O$$

$$2S + 3\ O_2 \rightarrow 2H_2SO_4$$

$$2KSCN + 4\ O_2 + 4H_2O \rightarrow (NH_4)_2SO_4 + K_2SO_4 + 2CO_2$$

Thiobacillus ferrooxidans derive energy from the oxidation of ferrous iron as follows :

(Pyrite) $2FeS_2 + 2H_2O + 7O_2 \rightarrow 2FeSO_4 + 2H_2SO_4$ (Mostly chemical)

$4FeSO_4 + O_2 + H_2SO_4 \rightarrow 2\ Fe(SO_4)_3 + 2\ H_2O$ *T-ferroxidans*

$2S + 3O_2 + 2H_2O \rightarrow 2\ H_2SO_4$ *T-thiooxidans*

The mechanisms can be summarized as under

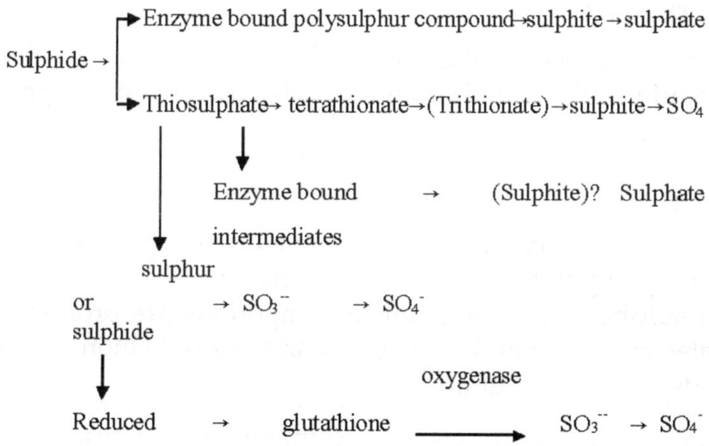

The biochemical a Pathway can be summarized as

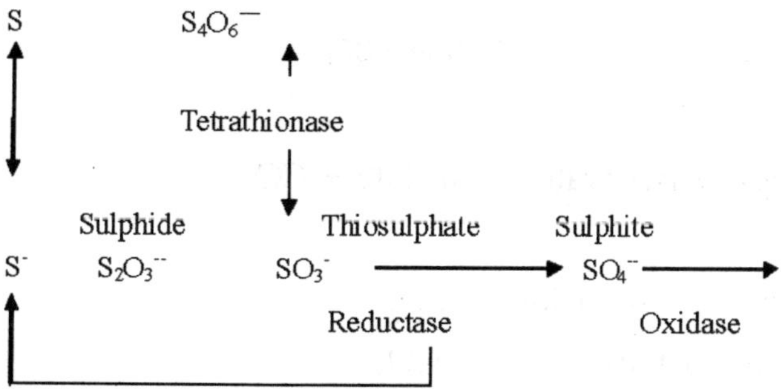

Adavantages of S-oxidation

Microbial sulphur oxidation is beneficial to soil fertility.

1. Formation of SO_4^{-2}as plants available nutrient.
2. Acidity is produced in the form of H_2SO_4 which solubilises many and reduces solicity and salinity.

Factors affecting S-oxidation

Temperature, soil pH, moisture, soil structure etc.

The oxidation of S is also carried out by many heterotrophic microorganisms aeorobic bacteria, actinomycetes and filamentous fungi. These organisms fall into two categories.

1. Oxidising elemental S and thiosulphate to polythionates.
2. Converting polythionates to sulphate.

Sulphate Reduction

SO_4^{-2} reduction takes place particularly in waterlogged, flooded rice soil. Under well aerated conditions this process is of insignificant amount.

Organisms

Dissimilatory sulphate reduction is the proper of a few anaerobic bacteria and many heterotrophic organisms where by sulphides and elemental sulphur (S) or inorganic S compounds are produced H_2S accumulates at the expense organisms and the oxidation of organic compounds they carry out are :

1. *Desulfovibrio desulfuricans* (most common and important bacteria for this process).

   → Lactate → Acetate + CO_2

2. *Desulfotomaculum*
3. *Desulfomonas* → pyruvate → Acetate + CO_2
4. *Clostridium*
5. *Desulfococcus* → fatty acids → CO_2
6. *Desulfosarcina* → fatty acids → CO_2

7. *Desulfonema* → fatty acids CO_2
8. *Desulfobacter* → Acetate → CO_2
9. *Desulfobulbus* → Propionate or lactate → Acetate + CO_2

In the SO_4^{-2} reduction the SO_4^{-2} acts as the terminal acceptor of electrons in the reduction of these bacteria. Bacteria like No.1 make use of sulphite, sulphate, thiosulphate, tetrathionate as e^- acceptors, but not oxygen nor organic sulphur compounds.

Biochemistry of Sulphate Reduction

During the reduction of SO_4^{-2} there is simultaneous oxidation of many organic compounds like carbohydrates, organic acids, amino acids acetate, pyruvate, lactate, fatty acids etc.

Two typical reaction for *D. desulfuricans* are given below :-
(Acetate is a common product)

$$4CH_3.CO.COOH + Na_2SO_4 \rightarrow 4\ CH_3\ COOH + 4CO_2 + Na_2S$$

Pyruvate Acetate

$$2CH_3CHOH\ COOH + Na_2SO_4 \rightarrow 2CH_3COOH + 2CO_2 + Na_2S + 2H_2O$$

Hydrogen and are therefore capable of autotrophic growth.

$$4\ H_2 + CaSO_4 \rightarrow H_2S + Ca(OH)_2 + 2H_2O$$

The first step in the reduction process is activation of sulphate by esterification to an aderyl group by the anzyme ATP sulfurylase to form adenosyl phosphosulphate (APS) and pyrophosphate. The APS is reductively cleaved to yield AMP and bisulphate which is sequentially reduced to trithionite, thiosulphate & sulphide.

The enzyme bisulphate reductase is present in all sulphate reducting bacteria, catalyzing reduction of bisulphate to trithionite.

The pathway can be summarized as :

SO_4^-ATP ↔ Adenosyl Phosphosulfate (APS) + PP → 2P

HSO_3^-(bisulphate) + AMP

↓$2e^-$

$S_3O_6^-$(Trithionate)

↓2e-

$S_2O_3^-$(Thiosulphate)

↓2e

S^-(sulphide) (H_2S etc.)

Condition for SO_4 Reduction

1. Anaeorobic condition are the primary requirement for SO_4 reduction. Soon after flooding oxygen disappears in the following

   NO_3^- - NO_2^- - Mn^{+++} Compounds→ Fe^{+++} - SO_4^-

   (Manganic) (Ferric)

   Sulphate is not reduced to S^- if oxygen and NO_3 are present.

2. The SO_4^{-2} reducing bacteria are very salt tolerant and function in pH range of 5.5 - 9.0
3. Heavy textured soil, submerged soil etc.
4. Addition of organic matter with wide C : S ratio.

Accumulation of H_2S is toxic to the plants and activities of *Desulfovibrio* also cause tremendous economic losses to the industry.

Microbial Transformation of Phosphorus

Micro organisms brings about a number of transformations of the elements - viz:

1. Solubilising inorganic compound of phosphorus.
2. Immobilisation of available P.
3. Mineralisation of organic P.
4. Oxidation or reduction of inorganic P compound.

Source of P in soil

Organic P : Agricultural crops contain about 0.05-0.50% P in their tissues, the compounds containing pure phytin, phospholipids, nucleic acids, phosphorylated sugars, coenzymes and related compounds.

Inorganic P : Mineral forms $H_2PO_4^-$, HPO_4^{-2}, apatities Fe and Al phosphates.

P-Cycle

The transformation can be illustrated in a cyclic form as:

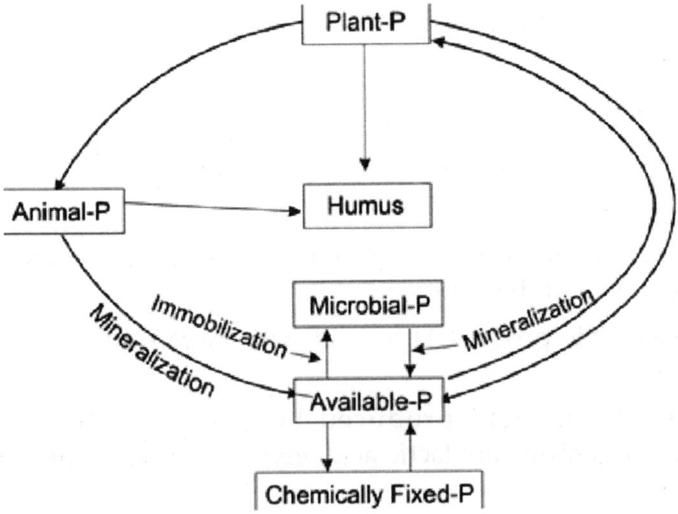

Living forms of P

Unlike NO_3^- sulphate which are assimilated by the plants only in their reduced forms [- NH_2 (amino) and -SH (sulphydryl)]. Plants do not reduce phosphate as is evident from the structure of phytic acid and lecithin (Phospholipid).

$H_2 - C - O - R$

|

$H - C - O - R$

O

||

$H_2 - C - O - P - O - CH_2 - N - (CH_3)_3$

|

O^-

Lecithin - (R = Fatty acids)

In soil of the total P 15-85% is in organic forms.

Solubilization of Inorganic – P

Insoluble inorganic P forms like $Ca_3(PO_4)_2$ etc. are converted by solubilisation action of many species of:

Pseudomonas, Mycobacterium, Micrococcus, Bacillus, Flavobacterium, Penicillium, Selerotium, Fusarium, Aspergillus etc.

1. No only Ca by Fe, Al, Mg, Mn and other forms of P are also solubilised.

   The solubilisation is brought about by the release of organic acids. The NH_4^+ and sulphur oxidizing chemoautotrph release HNO_3 and H_2SO_4 solubilising Ca, Fe, Al etc. phosphates.
2. In flooded conditions iron in insoluble ferric phosphates is converted (reduced) to Fe^{++} ferrous and concomitantly release phosphate in the solution.
3. Anaeorobic product H_2S react with Fe_3PO_4 to yield Fe_2S and H_2PO_4.
4. The organic acid produced by the metabolic activity of microorganisms are lactic acid, glycolic, citric, formic and acetic acids.
5. 2-Ketogluconic acid and similar acid are very effective chelating agent capable of complexing with Ca, Cu, Ni, Mn, Fe, and Al salts of phosphates with resultant solubilisation of phosphates.

Mineralisation of Organic Phosphorus

Conversion of organic forms of P into available inorganic forms is called mineralization.

Organisms

All microorganisms requiring P for their life cycle specially the heterotrophs make the bound element in the remains of the vegetation and in soil organic matter available to the succeeding plants.

Species of *Aspergillus, Penicillium, Rhizopus, Cunnighamella, Arthrobacter, Streptomyces, Pseudomonas* and *Bacillus* can mineralize the organic P. All these organisms possess phosphates and Phytase enzymes. Many types of fungi specially mycorrhizal fungi and actinomycetes also carry out mineralization.

Condition

1. More rapid in virgin soil.
2. Decomposition favoured by warm temperature thermophilic range (30° C to 50°C) more favourable than mesophilic range (<20°C)
3. The rate increases with rise in pH (acidity to neutrality increases phosphatase activity).

4. quantity and type of substrate, soil rich in organic P will result in more mineralization of P.
5. Bacterial cell P is mineralized quickly and its acid soluble organic P, phospholipids and DNA are dephosphorylated in short period.
6. P in microbial RNA is released more slowly.
7. All condition favouring ammonification favours P mineralization.
8. A C : P ratio of <200 : 1 will result in net P mineralization while if it is 300 : 1 there will be immobilization.

Mechanism

Highly significant correlation exists between mineralization of N and P. The amount of N mineralized is 8-15 times more than that of P mineralized.

There are many enzymes that cleave P form organic substrates. They are collectively called phosphatases. A general reaction is:

O   O

| | | |

R - O - P - O - H + H_2O ! ROH + HO - P - OH

| |

OH   OH

R - May have numerous structure.

Phosphatase catalyze cleavage of ethyl phosphate, glyceroph-osphate, phenyl phosphate. The substrates with two -R groups. Diesters will have different enzymes in contrast to single R groups (monoesters).

In soil presence of inositol phosphates with 5 or 6 or less P atoms per molecule suggests that a breakdown of the inositol hexaphosphate takes place. The enzyme phytase 1 : berate P form phytic acid or its Ca or salt (phytin) with resultant accumulation of inositol.

In general phytin derivatives are more resistant to mineralization hence phytase is more stronger then phosphatases.

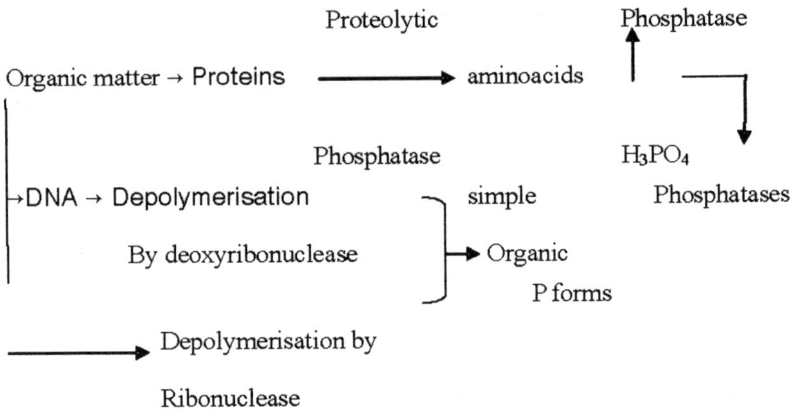

Immobilisation

Conversion of available forms of phosphorus into organic form by microbial activity is called immobilization of phosphorus.

Almost all kinds of microbes assimilate P in nucleic acid, phospholipids, other protoplasmic substances.

Organisms

1. The aeorobic organisms possess great capacity to immobilize P as compared to anaeorobic population.
2. The capacity of P immobilization of various organisms is in the order fungi > bacteria > astinormycetes.
3. *Aspergillus niger, Azatobacter* etc. are very sensitive for adequate P supply.

Conditions

1. Incorporation of organic residues with a wide C:P ratio (>300:1)
2. Well aerated normally moistened soil.
3. Medium textured soil.
4. Incorporation of animal residues possess less susceptibility of P immobilization.

Usually immobilization of N entails immobilization of P as well.

The organic P : organic N ratio generally remains 10 : 1 hence the C : N : P ratio 100 : 10 : 1.

Oxidation-Reduction Reaction

Oxidation

P like N exist in a number of oxidation states ranging form -3 of phosphene (PH_3) to oxidized state +5 of orthophosphate. Many of these redox transformation of P are biologically carried out.

e.g. $HPO_3^{-2} + ½O_2 \rightarrow HPO_4^{-}$

Phosphite (reduced form of P)

This conversion is brought about microbiologically as is evident that on addition of a biological inhibitor such as toluene this reaction is eliminated. The process is carried out by unspecified microbes only.

A number of heterotrophic bacteria fungi and actinomycetes utilize phosphate as their P source in culture media.

Reduction

When there is an anaeorobic condition and there is no supply of NO_3^- or SO_4^- then phosphate acts as the terminal e^- acceptor.

$$\underset{(+5)}{H_3PO_4} \xrightarrow{2H} \underset{(+3)}{H_3PO_3} \xrightarrow{2H} \underset{+1}{H_3PO_2} \xrightarrow{4H} \underset{-3}{PH_3} \text{ (phosphen)}$$

Phosphite Hypophosphite oxidation State

The quantitative significance of such biological reduction of P even in logged water soil is of practical value only.

Organisms like *Clostridium butyricum* and *Echerichia* coli are known to bring about this reduction.

□□□

Oxidation-Reduction Reactions

Oxidation

Chapter-6

Soil Biota and Organic Farming

Significance of Soil biota in Organic farming

The ill effects of long usage of chemical fertilizers and pesticides have resulted in rendering the soil less fertile leading to the depletion of vital nutrients which is a serious threat to ecology and environment. The only option left with us is, ecofriendly natural farming or ecofriendly organic fanning where the proper exploitation and utilization of naturally available resources is possible. Among many natural resources for ecofriendly organic farming, soil forms the important medium for the survival, growth, multiplication of wide variety of soil invertebrates most of which are directly or indirectly involved in enhancing soil fertility and growth of the plant The ecofriendly inputs for organic farming include the use of compost, FYM, green manure, biofertilizers, vermicompost, biopesticides, biocontrol agents etc. Therefore, the organic resource management is the backbone of organic farming and sustainable agriculture.

Soil is a living heterogeneous system with characteristics of physical, chemical and biological properties. Soil is inhabiting a vast group of soil biota including most of the vertebrates by which they make significant contributions to the soil fertility and crop production. The awareness on the existence of life process in the soil was aroused with the publication of Charles Darwin's classical book "The formation of vegetable mould through the action of worms" in 1881. During the same period, Muller (1879) recognized the importance of soil invertebrates in the establishment of different humus forms. These soil invertebrates can be manipulated to improve the soil physical, chemical and biological properties and regulate the decomposition processes to release the vital nutrients for plant growth. These are broadly categorized into soil microorganisms, nematodes, earthworms, soil micro and macro arthropods which directly or indirectly are contributing to the soil ecosystem.

Soil Microorganisms

The fundamental characteristic of soil is that it has its capacity to produce green plants. Plant and their products after being utilized by man and animal are thrown away as waste material to the soil. Fortunately microorganisms act upon and convert them into useful or harmful substances. In the words of noble laureate Waksman "Without microbes the life on this planet would come to a standstill". The different kinds of microbes such as bacteria, fungi, actinomycetes, algae, protozoa and viruses present in soil bring about several changes.

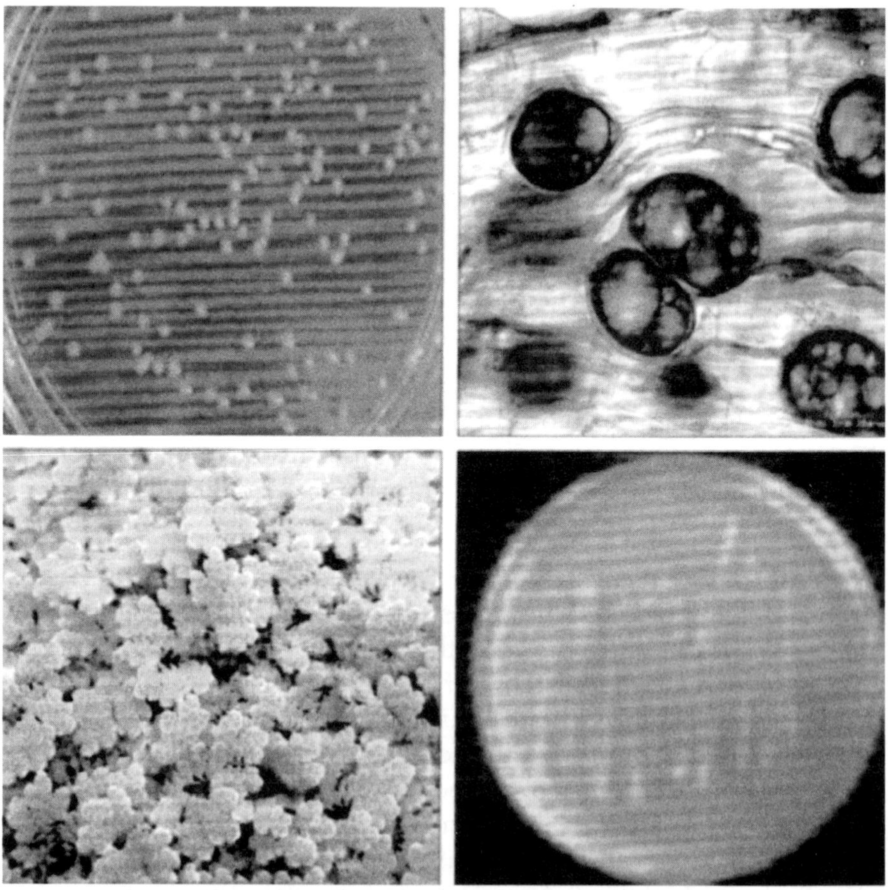

Fig. 6.1 : Soil Microorganisms

Bacteria

These are the most dominant group of soil biota, their population ranges from hundred thousand to several hundred million per gram of soil depending upon physical, chemical and biological conditions of soil. They live in soil as cocci, bacilli or spirilli. Some of the bacilli persist in unfavourable conditions and great resistance to prolonged desiccations and also to high temperature as endospores.

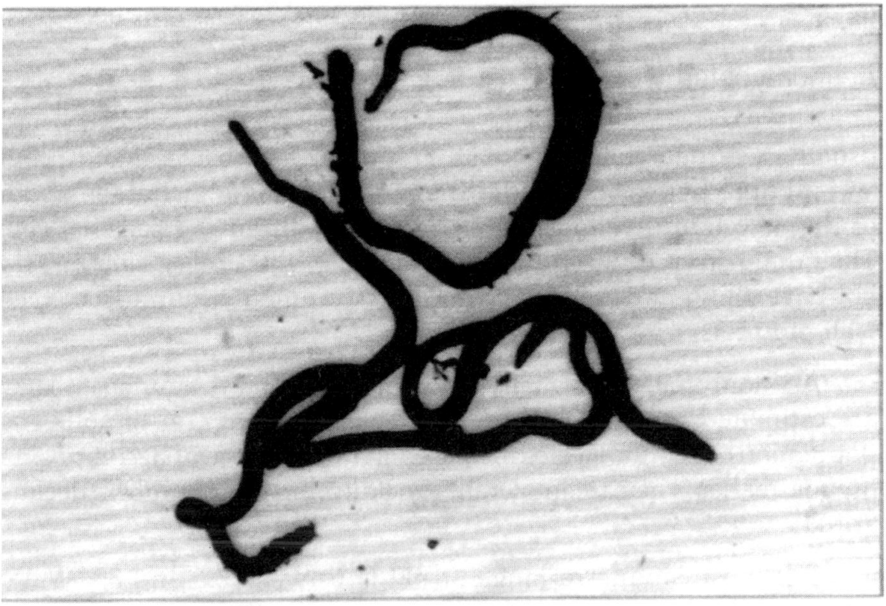

Fig. 6.2 : Earthworms for vermiculture

Most of the soil bacteria are heterotrophs and saprophvtic in nature. They help in decomposing organic matter and involve in weathering and decomposition of rocks and minerals which help in soil formation processes. Most of the bacteria secrete extra polymeric substances which help in binding soil particles, improve soil structure and texture for improved soil porosity for better exchange and infiltration of gases. Bacteria present in soil also involves various Geo-bio-chemical cycles like nitrogen, sulphur and carbon cycle etc.

Many beneficial bacteria present in soil has the capacity to use atmospheric nitrogen as biological nitrogen fixers. *Rhizobium, Azotobacter, Beijerinchia, Azospirillum* are the common bacteria present in this category Some of the bacteria like Bacillus., *Pseudomonas* sp., and fungi like *Aspergillus, Penicillium, Fusarium* present in soil help in solubalization of unavailable or fixed form of phosphorous present in soil. Of late *Pseudomonas fluorescens* has been receiving greater attention all over the world for the biocontrol of different crop diseases. The ability to grow in the absence of oxygen is an important biochemical trait which helps in the recognition of two broad categories of bacteria as aerobes and anaerobes. Under anaerobic conditions bacteria dominate the scene and carry out the microbiological activities in the soil since other organisms usually do not grow well in the absence of oxygen.

Actinomycetes

These are numerous and widely distributed in soil. They are usually regarded as intermediate group between bacteria and fungi and produce hyphae or conidia or sporangia like fungi, but the width of the mycelium is almost similar to the width of the bacterial cell. They differ from fungi in the composition of their cell wall. They do not have ehitin and cellulose, which are commonly found in the cell wall of fungi.

Actinomycetes are usually second to bacteria in abundance. They are estimated to be ranging from 10^4 to 10^8 per gram of soil. The population of actinomycetes increased in the presence of decomposing organic matter. They are intolerant to acidity and their numbers decline at pH5.

The most conducive range of pH is between 6.5 and 80. Water logging soil is unfavourable for the growth of actinomycetes whereas desertic soils of arid and semi-arid zones sustain sizeable population, probably due to resistance of spores to desiccation. The percentage of actinomycetes in the total microbial population increases with the depth of the soil and these can be isolated in sufficient number even from soil samples obtained from the C horizon of soil profile. Temperatures between 25°C and 30°C are conducive for the growth of aetinomycetes although thermophillic species growing at 55°C and 65°C are common in compost heaps where they are numerically extensive and mostly belong to genera Thermoactinomyces and Streptomyces. Many species of Streptomyces are known to produce antibiotics. Actinomycetes also play a role in the decomposition of resistance components of organic matter and in the formation of humus. Thermophillic actinomycetes dominate and actively participate in the transformations taking place in the compost pits, manure pits and hay.

Fungi

These are present in soil as mycelial hits or rhizomorphs or various types of spore forms. Their number in the soil varies from few thousand to few million per gram. A wide range of soils under various vegetation types and geographical areas have been examined for the occurrence of fungi. *Aspergillus, Mucor, Penicillium, Trichoderma, Cladosporium, Verticillium, Rhizopus, Fusarium, Cephaloporium, Botrytis* etc. are the most commonly encountered genera of fungi in the soil.

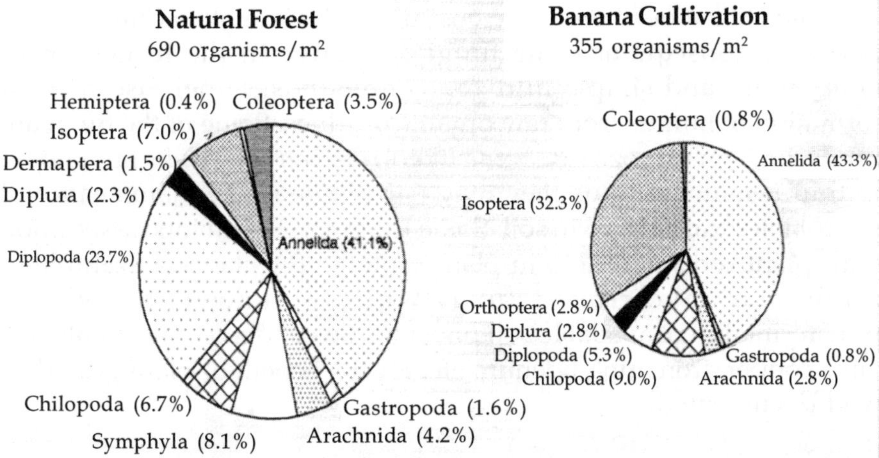

Fig. 6.3 : Comparision of density and diversity of soil macro-arthropod between natural forest and banana cultivated soils

Fungi being aerobic are more numerous in surface layers of well aerated cultivated soil. The quality and quantity of organic matter present in soil has direct hearing on fungal population, since majority of fungi are heterotrophic in nutrition. Fungi are dominant in acid soils compared to neutral and alkaline soils.

Filamentous fungi play an important role in the degradation of cellulose, hemicellulose, starch, pectin and lignin which arc present in organic matter added to soil. It is implicated that some fungi like *Trichoderma* sp., *Aspergillus* sp., *Penicillium* sp. and *Pleurotus* sp acts as most efficient organic matter degraders. The genus *Trichoderma* is a source of promising biocontrol agent against many plant pathogens. They grow rapidly on substrates and a wide range of fungal pathogens. *Trichoderma harzianum, T. viride, T. koinigii* and *T. virens* have been reported to control many soil borne diseases such as wilt, root rot, collar rot etc in many field crops. In addition TT *viride* is known to produce toxic chemical as viridian which prevents egg hatching and kills the larvae of root knot nematodes by rupturing their cuticle. Besides they are very good plant growth promoters, *p solubalizers* and organic matter decomposers. These species have been practically tested under field conditions and formulated and recommended for integrated disease management including seed treatment to many field crops, especially in organic farming. Many organizations have taken up the large scale production of *Trichoderma sp.* and thus they have become commercially important as biocontrol agents.

The utilization of proteinacious substances is another characteristic and as a consequence, the fungi are active in the formation of ammonium and simple nitrogen compounds. Fungi also help in formation of humus from raw organic residues. Some of the fungi are predacious on protozoans. Some like *Arthrobotrys conoides* are known to trap nematodes, thus they may participate in the maintenance of microbiological balance in soil. Some fungi form symbiotic association with plant roots and help in plant growth. The classical example of arhuscular mycorrhizal fungi which is known to help in transformation of available soil phosphorous and micro nutrients like zinc, copper, iron, molybdenum etc. for the benefit of plant growth is well documented.

Algae

These are found in different soils. They are abundant in places with enough moisture and sunlight. Their number, usually ranges from 100 to 10,000 per gram of soil. They form a scum on the surface of the soil visible to the naked eye. They are unicellular or filamentous or colonial. Green algae, blue green algae and diatoms are numerous in soil. By virtue of the presence of photosynthetic pigments of their cells, algae are phototrophic and use carbon dioxide from atmosphere and give out oxygen. Green algae are dominant algal flora in acid soils hut still numerous in neutral and alkaline soils. *Chlorella, Chlamydomonas, Chrocococcum, Oedogonium* are the most common green algae found in the soil. Blue green algae are pigmented having phycocyanin in addition to chlorophyll. They do not posses flagella and do not produce sexually. Some of the blue green algae produce specialized cells known as heterocyst which are implicated in nitrogen fixation. The waterlogged rice soils provide an ideal environment for the growth and multiphcation of blue green algae. Blue green algae also add lots of organic matter to soils and improve soil aggregation.

They also grow on eroded and abandoned soils and also volcanic deposits when the required moisture is available. The mucilage of these algae has the properties of absorbing water quickly and retaining it for long time. Thus the growth of these algae provides a substratum for the growth of spores, grass seeds etc and thus help in building soil fertility The thick growth of such algae on the surface of such soil acts as surface binders and prevents soil erosion.

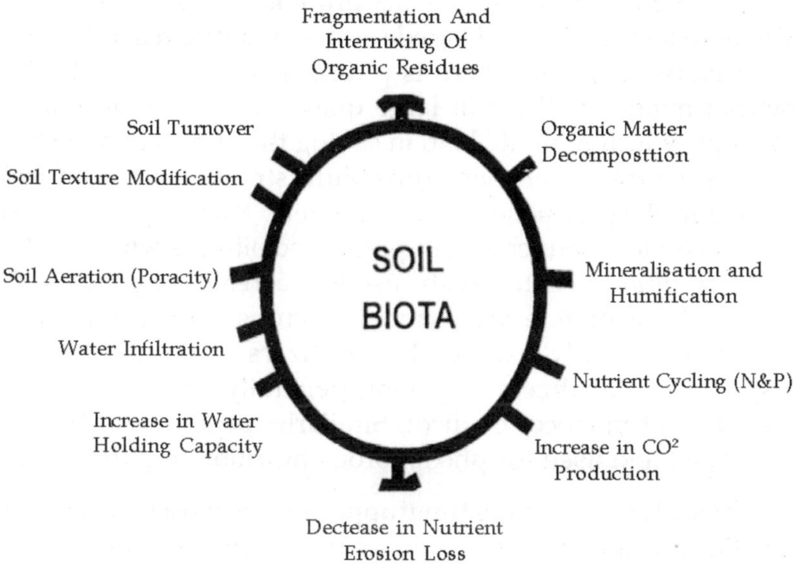

Fig. 6.4 : Integrated activity of soil biota as reflected on physico-chemical characteristics of the soil

Protozoans

These are also common in soil where, the population ranges from 10,000 to 100,000 per gram of soil. They are characterized by a cyst stage in their lifecycle which can help the species to withstand adverse soil conditions. Most of the soils contain protozoans like flagellates, amoebae or ciliates. Flagellated protozoan are predominated as compared to Amoebae and Ciliata. They are abundant in upper layer of soil. Application of organic manures increases the population of protozoans which is a reflection on the corresponding increase in the bacterial flora due to the addition of organic mater. Soil moisture level is another important factor affecting soil protozoa hut they could survive even during the absence of moisture by forming cysts especially in dry condition.

Viruses

These are sub microscopic and obligated on other soil microorganisms like bacteria, fungi, actinomycetes and viruses. Some of the plant viruses are said to be soilborne, they do not multiply in soil hut persist for some time as obligate parasites. There is no strong evidence to conclude that they have any relationship to enhance the crop yields or their overall influence of soil on agricultural productivity.

The beneficial microorganisms present in soil will generally be low in numbers and hence desired results cannot be reaped from these microorganisms. Hence there is a dire necessity to produce these beneficial microorganisms in large quantities for application to the field crops, which not only help in saving the application of chemical fertilizers. Currently efficient rhizobium strains have been isolated and used in the production of biofertilizers. *Rhizobium* is extensively used in dicoryledonous crops like pulses and oilseeds where *Azotobacter* and *Azospirillum* is extensively used in cereals, vegetables, fruits, ornamentals, plantation and commercial crops. Azolla and blue green algae are commonly used as biofertilizers in paddy cultivation. Phosphorus solubalizers are being generally applied to all crops irrespective of monocot or dicot. Similarly, mycorrhizal fungi were also isolated and used for phosphorous nutrition in different crops.

Arbuscular mycorrhizal fungi application is most suited to nursery crops. But the quality of the produce was improved due to secretion of growth promoting hormones like Indole acetic acid, Gibberlic acid, Cytokinins and vitamins that help in deterring root borne pathogens.

These beneficial microbial inoculants are mass multiplicated in large quantities on appropriate medium on large scale fermentors and will be mixed in pre sterilizes carrier material like lignite, wood charcoal, talk exfoliated vermiculate, peat etc and are packed in materiel like low density standard polythene bags. Arbuscular mycorrhizal is an obligate parasite and can be mass multiplicated in the presence of hosts like blue panic or guinea grass. In recent years big commercial organizations like KRIBI-ICO, MFL, and IFECO etc have ventured into this business for large scale production and marketing. These microorganisms will survive for long time, are ecofriendly in nature and help for soil fertility sustenance for long time. Thus their contribution is immense in the field of organic farming.

Nematodes

Incredible nematodes live in soil and participate in the food webs at several trophic levels. Soil inhabiting nematodes are classified as free living bacterial feeders, fungal feeders and predators. One entomopathogenic nematode, *Steinernema carpocapsae* has been identified as an effective biocontrol for wide range of pests including soil pests like rootgrubs. A bacterium, *Xenorhabdus sp.* is symbiotically associated with this nematode which actually kills the pests. This nematode is commercially available for field application. Predacious nematodes such as *Seinura sp.* and *Strongylids sp.* are known to predate on root feeding nematodes, *Meloidogyne sp.* soil protozoa, bacteria etc. Free living nematodes are saprophytic in nature and feed on organic matter which helps in decomposition and indirectly participating in soil fertility but not so significant like other soil invertebrates such as earthworms, microorganisms and micro-arthropods.

Earthworms

Earthworms are the important soil invertebrates have an widespread distribution and show a high degree of adoptability to the prevailing soil conditions. Their involvement in the soil ecological process contributing to the soil fertility was first recognized by Charles Darwin (1809-1884) and called the "Earthworms as farmers friend" and also stated that "these lowly organized creatures are so important as they played an important role in the history of the world. This statement is so relevant in the present scenario of organic farming and sustainable agriculture.

Globally more than 7000 species of earthworms has been distributed in all the habitats except cold and desert regions. Ecologically they are classified as 1) Epigeic earthworms confining to the topsoil feeding on the organic matter at surface feeding zones. 2) Endogeic earthworms are completely soil dwelling feeding on the organic matter at subterranean feeding zone. 3) Anecic earthworms are deep soil inhabitants making horizontal and vertical tunnels for feeding activities. The Epigcic earthworm species are most important as surface feeders on organic matter include *Eudrilus eugeniae, Perionyx excavatus, P. sansibaricus, Eisenia foetida* and *Pheretima* sp. And they can also be used for vermicomposting.

Vermiculture is the art of maintaining and multiplication of earthworms in semi-arid conditions for the purpose of vermicomposting. The most preferred and generally used species of earthworms include the African night crawlers, *Eudrilus eugeniae*, the banded or tiger worm, *Eisenia foetida* and *Perionyx excavatus*. It is always good to use all these earthworm species in combination so as to get enhanced microbial density and reflect on the quality of vermicompost.

Biology

The body of earthworm is longitudinal and segmented. The colour is more intense on the upper surface because of the presence of pigments usually reddish brown or red in the cells of skin but much of the colour is due to haemoglobin in the earthworm. They have no special breathing organs, oxygen is absorbed and waste gases are eliminated by diffusion through the skin from a network of fine blood vessels just beneath the skin surface. Gland cells in the skin secrete mucus to keep the surface moist and lubricate the body as it moves through the soil that is how the earthworms are moist and slippery to touch and because they must constantly replace water lost by evaporation from the surface and they cannot live in very dry soils.

Earthworms have both male and female reproductive organs in the same individual as hermophrodites with two pairs of testis and one pair of ovaries. During mating two earthworms align themselves in head to tail attitude with their lower surface touching each other over a length of about 35 anterior segments. A thick layer of mucus is produced covering each worm from near the front end to a few segments behind the clitellum. Sperm celis are discharged by both worms, the sperms from each worm find their way to the sperniathecae of each other. Sperm cells are released to fertilize the eggs and forms the oval egg case or cocoon. The cocoons are deposited in the soil and fertilized eggs develop and grow into small worms taking 2-4 weeks to hatch depending on temperature. The number of eggs/cocoon is variable, but usually only one or two worms succeed in developing and emerging from a cocoon.

Decomposition of organic matter

They prefer to live in a light, loamy soil and structure of heavy' clay soils should be modified for release of earthworm population. Horizontal and vertical burrows of earthworms provide channels for root, water and air penetration deep into the soil and also increase the volume of the soil from which plant roots can extract nutrients and water. The crumbs or aggregates of soil formed from worm casts are usually more stable than the hulk of the soil. Earthworms eat large quantities of soil as well as organic materials. Everyday they eat about half of their own weight of soil.

Species that excrete their faeces as casts on the surface of soil have been shown to bury the original surface to a depth of 4-5 cm. the common earthworms in any garden soil tend to excrete their faeces underground.

Earthworms breakup, eat and mix the large amount of organic material in to the soil. Individual earthworm can eat as much as 40 gram div matter per year. This works out at 80 tonnes/ha/year for a typical population of 200 worms/sq. m. This number may be reduced due to longer dry period when earthworms are inactive.

Earthworms eat to live but an important benefit for soil and plants is that organic materials are broken up and mixed with soil. In the process, the nutrients in them become much more readily available to soil microorganisms and to plants. The increased microbial activity further aids organic matter decomposition and hence nutrient available to plants. Studies elsewhere have shown that about 6% of the nitrogen in organic matter previously unavailable to plants becomes available after passing through digestive tracts of earthworms. Calcium, Magnesium, Potassium, Phosphorus and Molybdenum are many times readily available in worm casts than in the surrounding soil and casts are also nearly neutral pH than the soil. Increased availability of nutrients is one reason why crop yields are generally increased by earthworms.

Observations have indicated char some soils have 20% greater water holding capacity since the introduction of some species of earthworms. This means that more water is available for plant growth. Field experiment conducted elsewhere to determine the introduction of earthworm *Pontoscolex coretyhrurus* at the rate of 360 kg/ha for soil fertility and crop production by cultivating 4 crops; i.e. maize followed by rice, cowpea and rice. Nitrogen and carbon contents were recorded 18 and 7% higher in earthworm treated crop fields. The establishment of earthworm population had seen in all the subsequent crops also.

Culture methods

The Vermiculture units could be as simple as eras cardboard boxes or even cement or stone slabs tanks. Any container with 25 to 30 cm deep similar to bath tubs like troughs or drums can be used for small scale vermicompost production. Permanent beds with cement concrete tanks also with the higher dimension of 4 m length × 1.5 m width × 0.75 m height. Such tanks are recommended to use in organic farming villages for vermicomposting. The temperature and moisture requirements of earthworms need to be kept in mind while selecting the sites for vermicomposting tanks. Earthen pits with still longer dimensions are also used in most of the sugar factories where sugarcane trash is being utilized for making vermicomposting. The worms are best added to the medium from which they are collected. The best should be kept moist her not saturated with water. Placing the wet gunny bags on the surface reduce evaporation losses and provide darkness to earthworms. Frequent monitoring is required to make sure that the bed is moist enough so that the worms are healthy with continuous breeding activity.

The decomposed organic leafs' materials along with cattle dung and moisture of about 40-60 % is essential for Vermiculrure. The units can be mulched with gunny bags or leafy materials. One kg of earthworms containing about 1000- 1200 numbers can yield 2000 cocoons/week and 4000 young ones emerge and from this 7000 to 8000 earthworms can be produced in about 3 months. Thus the culturing of earthworms is more profitable and a kg of earthworms cost around rupees 450 to 600.

The availability of earthworm is the most critical factor to go for vermicompost production which is universally known as vermicomposting. It is very simple and now it has become popular technology among the farming community in India.

Vermicomposting can be taken up by anybody including farm men, women, urbanites, terrace gardeners etc. The main principle behind that any biodegradable organic waste including agro based or city solid waste or biodegradable industrial waste (excluding plastic material) can be used for vermicomposting. The agro based waste need to be decomposed by mixing with one part of cattle dung with three parts of biodegradable wastes. The decomposition process will continue depending on the C: N ratio of organic materials used and essential moisture between 40 and 60% during which frequent turning or mixing has to be done to hasten the decomposition process and may need about 20 to 30 days for decomposition.

The size of tank is individual choice depending on the amount of waste available and only height of the tank is constant, i.e. 1 m for all different dimensional units, the right time to release the earthworms into these tanks when the decomposition of organic material is completed atleast 70-80%. About one kg of earthworms (1000-1200 worms) are required for the tank size of one sq. m. The earthworms consume the decomposed organic waste and convert in to vermicasts to form vermicompost which can be harvested periodically once in 3-4 days. The composting process will be completed in 50-60 days and one can produce the vermicompost in 3 months period. The shelf-life of vermicompost is 10 to 12 months if it is preserved in cool and shady place. The vermicompost thus producef is a value added product since it has all the required major and minor nutrients and growth enhancers in addition to being 100% organic manure (Table 6.1). They should be taken to protect the units from sunlight and rainfall and also from common predators like ants, ground beetles, lizards, birds, rodents etc.

Advantages of vermicomposting

The Vermiculture and vermicompost technology offers many advantages to end users:

- Earthworms are the biological tools for healthy and productive soils and are decomposers of complex organic matter.
- Vermicompost can be used as effective fertilizer to almost all crops including vegetables, ornamentals and plantation crops.
- Vermicompost is an aerobically degraded organic manure which undergoes chemical changes by the enzymatic activity of earthworms and also associated microbial population.
- Vermicompost is a value addition product which is economically a viable venture.
- Vermicompost is the hot spot for multiplication of soil fungi and bacteria.
- Vermicompost technology is effective in urban solid waste management to reduce environmental pollution.

- The production of vermicompost is a highly lucrative business and income generating activity by way of marketing earthworms and vermicompost.
- Vermimeal is a dried extract of earthworms that can be used as feed for poultry birds, fishes and cattle.
- Vermiwash, the coelomic fluid extract of earthworms has medical properties such as antibacterial, antifungal and also pharmaceutical in medical field. It has also anti plant pathogen property which can be used as foliar spray to control the plant diseases.
- Vermicompost technology is a morale booster for farmers in organic farming and also farmers can avail loans from the nationalized banks to start commercial production of vermicompost.

Table 6.1 : The nutrient status of the vermicompost

Parameters	% or ppm
Organic carbon	9.15-17.98
Total nitrogen	0.5-1.5
Available phosphorous	0.1-0.3
Available potassium	0.15-9.56
Calcium and magnesium (meq/100g)	22.67-47.6
Copper	2.0-9.5
Iron	2.0-9.5
Zinc	5.7-11.5
Sulphur	128-548

Recommendations have been made on the use of vermicompost based on the feed back from the farmers who had higher yields in different crops. Application of vermicompost at 2 tonnes per acre in case of cereals, pulses and oilseeds, 4-6 tonnes per acre in case of vegetables, ornamentals and spices and in case of fruit crops, plantation crops like coconut and arecanut 2-5 kgs/tree depending on the size of the trees were recommended to the cultivators.

Earthworms play an important role in comminuting organic matter resources to produce value added vermicompost which can be used in organic farming. Hence, the earthworms are very optly referred as

"Cindrella of organic farming" for self sustained, ecologically safer and economically viable future.

Limitations

Earthworms are rich in natural rain forests where the temperature is optimum between 20° C to 30° C. Temperatures above 35° C under Indian conditions are usually fatal for earthworms and death is more rapid during drier periods. Some species remain dormant in cocoon stage to overcome the adverse conditions. Similarly earthworms do not tolerate too much alkalinity with more than 8 pH and too much acidic with less than 5 pH and the most optimum is 6 to 8 pH as neutral for better survival of earthworms in nature.

Earthworm's populations are reduced due to mechanical damage by agriculture implements like ploughs, spades, forks, power tillers and other implements. Experimental evidence is available to prove that man chemical pesticide sprays likc chlorinated hydrocarbons, Organophosphorus and carbamates are lethal to the earthworm populations. Application of fumigants such as chloropicrin, methyl bromide and DD for the control of soil inhabiting insect pests and nematodes wili reduce the earthworms populations. Similarly, the application of Arsenic and Copper compounds to control many plant diseases are also known to reduce the earthworm population.

Soil Microarthropods

The microarthropods generally found in soil measuring less than 10 mm in length can be considered as microarthropods including proturans, diplurans, collembolans, mites, isopods, pauropods, symphylans etc. they are distributed in all kinds of soil including organically rich grasslands, pasture lands, garden land, cultivated, uncultivated, barren land etc. The population of these microarthropods varies from place to place depending upon many factors such as soil moisture, organic matter content, cultivation practices etc. Prolonged inundation adversely affect the soil microarthropods population and the population increases with precipitation of organically rich soils. A total density of 16-20 thousands per sq m. has been reported from the soil containing higher organic mater content especially in the fhrest litter of Western Chats and Himalayas.

Among the soil microarthropods collembolan and mites are abundant and found to he predominant group of soil fauna that accounted for 78- 90 % of total arthropods fbund in soil. Collembalans are commonly called spring tails primitively wingless group and are found as abundant soil fauna. Soil mites including Crvptostigmata (orbatei) Mesostigmata and Astigmata surpass other soil arthropod group in density and species diversity Collembalans and soil mites are important detrivorous and fungivorous saprophagous etc. Proturans and Collembalans are very small, less than 2 mm in length primitively wingless, white, fragile insects. These are found to occur in organically rich soils mostly common during rainy season.

Similar to proturans and diplurans Pauropoda and Sympla are small, fragile, delicate, miniature group of Myriopoda that prefer to live in leaf litter of surface soils 'and they prefer to feed on decaying organic matter. Isopods are terrestrial crustaceans popularly known as woodlice associated with litter, logs, bark stones etc. They are known to live mostly in dark and damp places. They have ability to conserve moisture due to the presence of their water proof cuticle. Isopods mainly feed on dead and decomposing organic mater including faecal pellets of soil animals. Based on the field experiments conducted in different parts of India, soil faunal diversity are quite active under organic farming system in soybean, cowpea, ragi etc. during rainy season.

Higher abundance of oribatid mites, collembolans, mesostigmatids have been found to occur regularly in the plots applied with minimum of 15 tonnes per ha farm yard manure. These plots also recorded significantly higher organic carbon, available phosphorus, calcium and also enhanced physical properties such as pore space, bulk density, water holding capacity etc.

Application of 15 tonnes per ha of FYM helped to establish significantly higher population of invertebrates compared to chemical fertilizers alone. Similarly, partially decomposed FYM application helped to establish significantly higher population of soil invertebrates as compared to fertilizer application alone. Higher abundance of soil invertebrates such as oribatid mites, collembolans etc. had positive correlation with increased physico-chemical properties of the soil. Agricultural management practices such as indiscriminate use of fertilizers, pesticides, use of heavy implements, deforestation have adverse effect on the activity of soil microarthropods and their useful role in organic matter decomposition.

Soil Macroarthropods

The soil arthropods measuring more than 10 mm in length can be considered as macroarthropods including centipedes, millipedes, termites, ants, large number of ground dwelling beetles etc.

Millipedes and centipedes

Millipedes and centipedes are important soil macroarthropods found in arable and garden soil and are quite abundant during rainy season. Millipedes are generally recognized as feeders on dead and disintegrating plant material. They are also acting as agents for decomposition and enrich the soil fertility. They are comparatively longer in size (5-6 cm) and can be easily recognized by the presence of well developed dorsolateral expansions on all body segments. Millipedes are active burrowers in forest litter, compost pits and farm yard manure. Millipedes ingest the detritus along with soil and thereby mix the organic and inorganic components. Centipedes are distributed in moist habitats of tropical soils. They are often found under leaf litter, bark, rocks, cracks and crevices etc. They are carnivorous feeding on smaller insects which are associated in the leaf litter and smaller plants.

Termites

Termites are primarily cellulose feeders and are pests of many dry land crops and wood work of the buildings. Large population will be involved in burrowing the soil by moving in subterranean galleries to more than 100 m in search of food. Termites affect the soil properties through construction of mounds, nests, galleries and earthen runways on the ground. Termites wherever present play a significant role in soil modification and organic matter break down and their effects may be positive or negative for crop production. Termites activity may drastically influence the fertility process in natural ecosystems to the extent that the physical and chemical properties of soils may be largely determined by the digestive activities of the animals and modifications of the environment caused by them. According to Drummand (1886) that the termites are true analogues of earth worms and play vital role in turning over soil cover by mixing organic matter with the soil facilitating its decomposition. Since the termites are removing the soil from various depths and they arc more selective in clay, silt and fine sand particles to build their nests and earthen runways.

Termites build variety of mounds/nests either above ground or below ground with unilocular and multilocular arrangement of fungus gardens, these fungus gardens are sponge like structures and are built out of their faecal matter over which mycelium and conidospores grow for food and depositing the eggs and nursing the young ones. The excreta is also used in the construction of nest and lining the subterranean galleries and earthen sheeting's on the food surface. Fungus garden contains cellulose, lignin, carbon, nitrogen, mineral matter, fat in addition to smaller quantities of calcium, magnesium, sodium, zinc, iron, manganese and copper.

The mound soils are characterized by higher proportions of finer soil particles such as clay, silt and fine sand as well as organic matter, phosphorus, the exchangeable cations like Ca, Mg, K, and Na are recorded higher in the mound soils. Mound soils affect the vegetation by adding certain modified properties of the soil that influence the growth of plants as this was verified in many crops such as cowpea, ragi, sorghum, soybean, cotton, sugarcane, chilly etc. The growth of the plant was luxuriant, vigorous with larger pods and earheads and also higher yields on the mound soils compared to surrounding soil.

Ants

Ants are found everywhere on the earth marching in large numbers on the ground. Soil inhabiting ants exhibit wide variety of feeding habits such as omnivorous, carnivorous, fungivorous, grenivorous, detrivorous etc. Ants exhibit territoriality, good communication system, alarm and assembly, polymorphism, division of labour etc.

They build crater like nests on the ground by excavating the soils from deeper regions and as a result water percolation to the plants is found to be faster. Excavation of nests and deposition of heaps of soil ants hasten the decomposition of organic matter and also act as scavengers in removing dead insects from their habitats. These ants are closely associated with homopteran insects for feeding on honeydew, nectar of plants. They are predators to attack many harmful insects such as termites, cutworms, aphids etc. They are also house hold pests causing nuisance in the human habitations.

Beetles

Many beetles such as ground beetles, darkling beetles, tiger beetles and dung beetles are known to live in the soil during their immature stages. Among these ground beetles and tiger beetles are predators, darkling and dung beetles are nature's scavengers. Dung beetles mainly carry large quantity of dung under the ground and this precious organic manure is being preserved by these beetles naturally to enrich the soil. The larvae and adults are making tunnels by burrowing the soil and organically enrich the soil by improving the soil fertility in many cultivated crops.

Observations recorded on the development of soil macrofaunal communities and the corresponding effect on the soil physical and chemical properties as well as quantifying the role of these soil fauna in sustaining the soil productivity The diversity, density and biomass of soil faunal communities in the natural tropical forest are compared to those in plots of banana where the abundance of different macrofauna are, continued to exist under cultivation of banana.

Summarily the soil biota including microorganism, earthworms, micro and macro arthropods are playing very important and significant role in organic matter decomposition, mineralization by biochemical transformation to enhance the rapid recycling of nutrient and consistently make them available for plant growth. The natural resource management with recommended quantity of all the organic inputs is very much necessary to conserve the agro-ecosystem. There is a possibility of manipulating the soil biota to optimize its influence on soil fertility and plant growth. Therefore, soil biota is an important component for organic farming system and sustainable agriculture.

Good agricultural practices, such as conservation agriculture, can reduce the carbon footprint and the adverse environmental impacts of biofuel production-just as they can for extensive agricultural production in general.

In the last century, chemical fertilizers were used in agriculture. Farmers were happy of getting increased yield in agriculture in the beginning.

But slowly chemical fertilizers started displaying their ill-effects such as:

- Leaching out
- Polluting water bodies
- Destroying micro-organisms and friendly insects

- Making the crop more susceptible to the attack of diseases
- Reducing the soil fertility and thus causing irreparable damage to the overall system
- Biofertilizers are groups of live and latent cell of microbes or micro-organism which are used to inoculate seed, plant surface and soil or all three which increase the production or productivity without impairing soil fertility.
- The term bio fertilizer refers to preparation containing live or latent cells of efficient strains of microbes which helps in enhancing the soil fertility by
  - Fixing atmospheric nitrogen,
  - Solublization of phosphorus or
  - Decomposing organic wastes or
  - Augmenting plant growth by producing growth hormones with their biological activities.

Benefits from using biofertilizers

- Stimulate plant growth.
- Activate the soil biologically.
- Restore natural soil fertility.
- Provide protection against drought and some soil borne diseases.
- Increase crop yield by 20-30%. Cost effective.
- Supplements to fertilizers.
- Eco-friendly (Friendly with nature).
- Reduces the costs towards fertilizers use, especially regarding nitrogen and phosphorus.

Role of biofertilizers for sustainable agriculture

- Renewable source of nutrients.
- Sustain soil health.
- Supplement chemical fertilizers.
- Replace 25-30% chemical fertilizers.
- Increase the grain yields upto 10-40%.
- Decompose plant residues and stabilize C:N ratio of soil.

❐❐❐

Chapter-7

Agroforestry and Sustainable Agriculture

"Agroforestry" is the new name for an ancient landuse practice. The concept of Agroforestry implies the integration of farming with forestry practices on the farm to increase output. The practice of Agri-silviculture which was began way back in Burma in 1856 is a classical instance. Various tree species and annual agricultural crops can be grown together.

In our country, the functional allocation on land is 46.4 per cent for agriculture and 22.7 per cent area for forestry is not sufficient for meeting the multifarious requirements of growing human and cattle population for food, feed, fuel and other raw materials. The only answer appears to be to integrate the land use for agriculture and forestry in such a way as to maximise production of foods and services for diverse requirements for rural communities.

Modern agroforestry establishes a symbiosis among agricultural crops tree species and livestock raising. In other words, these are complementary and beneficial to each other.

Tree growth conserves soil moisture, increases atmospheric humidity, improves soil fertility, protects field crops against the scorching and desiccating effects of winds. Besides amelioration of adverse climatic conditions, trees supply cheap fuelwood in villages, so that cowdung, which is still being burnt on large scale (about 60 per cent) for cooking purpose, is released for use as farmyard manure. Fruit tree species enrich the diet of rural population and also meet the demand for timber to small scale rural industry. Fodder tree species provide the much needed top feed (fodder) for sustenance of livestock. In fact, there is a need to devise such a land management and farming system which would produce food, fruit, fodder and wood and at the same time conserve the ecosystem. Such a system is agroforestry.

Definition of Agroforestry

Agroforestry may be defined as "an efficient and integrated landuse management system by raising of certain agricultural crops, forest tree species and or animals simultaneously or sequentially on the same unit of land with appropriate managment practices which result in overall increase in the production under a particular set of climatic and edaphic conditions and socioeconomic status of local people". Agroforestry involves interaction of woody perennial ecologically and economically with the crop and or livestock.

Objectives of Agroforestry

The following are the main objectives of agroforestry :

- To manage land efficiently so that its productivity is increased and restored.
- To use available resources efficiently and economically.
- To generate employment opportunities to rural poor.
- To provide raw materials for developing small cottage industries in rural areas (raw materials namely, wood, pulp, fibre, medicinal material ingredients and oils and cakes, gum wax, resin, lac, tannins, dye, green manure, soap substitutes, perfumes etc.).
- To provide basic needs of small and marginal farmers for food, feed, fodder, fruits, firewood, small timber etc.

Higher sustained and more diversified productivity is obtained from the existing resources when positive interactions are at optimum levels.

Components of Agroforestry

There are three basic components of agroforestry.

- Agricultural crops/herbaceous plants
- Woody perennials (tree crops/forest plants)
- Animals.

Above basic components when combine together form three basic agroforestry systems.

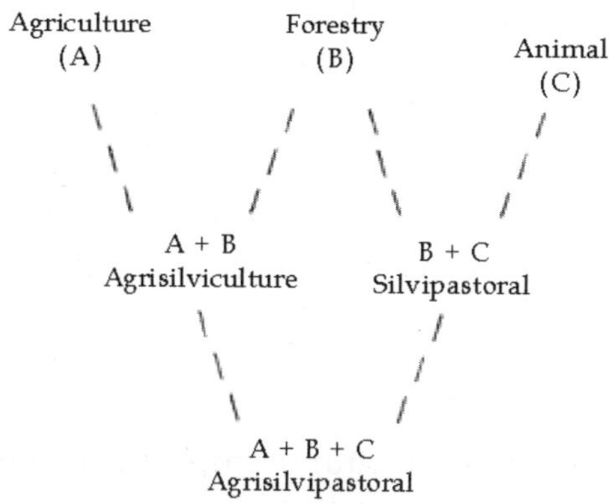

Forms of Agroforestry Practices

The following practices are being adopted by the farmers in India such as.

Trees with crops

This form of practice is very common in India where crops are grown with trees. This is achieved by five ways.

- Trees on crop land,
- Tree gardens,
- Taungya
- Improved fallows, and
- Plantation crop combination

Alley Cropping

It is an agroforestry technique of growing arable crops between hedge rows of trees and shrubs. Trees and shrubs in question are trained and pruned regularly in order to reduce vertical as well as horizontal growth this in turn will reduce the competition between trees and agricultural crops. Alley cropping has significant potential for maintaining soil fertility under continuous cropping. Trees and shrubs add organic matter and minerals through their leaves which is expected to improve physical, chemical and biological properties of soil. For alley cropping, *Leucaena* and *Glyricidia* species are most common.

Both the species are easy to establish and quick to regrow after pruning and lopping. Other crops namely pearlmillet and sorghum are reported to be most compatible for hedgerow intercropping with *Leucaena* wider alley width of 7.8 m is superior to 3.6 m from the point of yield of the arable crop. Similarly, cutting height of 15 cm is superior to 60 cm for better yields of both *Leucaena* and the arable crops.

Alley cropping is beneficial to the poor farmers and it provides greater total yield and a possible greater financial return resulting in a more uniform distribution of income.

Windbreak and shelterbelt

In order to prevent agricultural crops from intense sun, snow, and high wind velocity, the growing of trees or shrubs in rows is essential.

Inter and mixed croping

Agricultural crops and trees are intercropped and/or mixed cropped, e.g. maize and leguminous *Leucaena* species.

Kind of land and site for Agroforestry

The following type of land and sites can be assigned for agroforestry.

- Field boundaries
- Along with farm roads and nallah sites affected by erosion.
- Land on which cultivation is difficult.
- Old fallow lands.
- Cultivable wasteland.
- Site of pump houses, cattle sheds and kitchen gardens.

Agroforestry systems

The following are main sub-divisions or systems of agroforestry.

- Agri-silviculture
- Silvipastoral system
- Agrisilvipastoral system
- Horti-silviculture system

- Agri-horticulture system
- Agrihortisilviculture system
- Multipurpose forest tree production (other specialized agroforestry systems).
- Apiculture with trees
- Aquasilviculture or Aquaforestry
- Agrisilviaquaculture

Agri-silviculture

When agricultural crops are grown in combination with forest tree species, the system is known as "Agrisilviculture". For instance, when til, groundnut and arhar are grown as intercrop with *Leucaena leucocephala* and *Sesbania aegyptica,* This crop combination with forest trees results in an increase in both grain as well as fodder yields. In another combination like minor millet (*kharif*) and oats (*rabi*) has resulted in an increase in total yield when intercropped with *Leucaena.* Agri-silviculture system provides food, fuel, fodder, manure, paper pulp and packaging materials for the rural people.

There are number of ways to grow tree crops with agricultural crops :

- Tree species on the border of crop field or orchard.
- Tree species and agricultural crops in alternate rows.
- Tree species and agricultural crops in alternate strips.
- Tree species and agricultural crops in mixed form.

In most of the cases legume species are selected for agrisilviculture system.

Silvipastoral system

Silvipastoral system is a land management system, wherein the inter-spaces between two rows of forest tree species are utilized for cultivation of grasses and grass legume mixtures. This system provides fodder, fuelwood and small timber.

Land which is not suitable for raising agricultural crops should be developed as pasture and multipurpose tree species should be planted. These trees will provide pods and leaves for stall feeding during the leanperiod. It is advisable to plant such species at wider interval least their canopies suppress the grass cover. Planting of trees on pasture land improves the micro-climatic conditions like regulation of wind flow over the area, shade and shelter to the grazing animals'61. Under ideal silvipastoral system, there is non-competitive utilization of environmental resources. This requires optimization of biomorphs that constitute the silvipasture.

Under arid conditions of Rajasthan, *Sesbania grandiflora* and *S. sesben* increase forage production of *Cenchrus ciliaris, Setaria anceps, Desmanthus Chrysopogon fulvus* gives higher yield when it is grown with *Eucalyptus hybrid.*

Silvipastoral system provides '**two tier grazing**' under *in situ*. During rainy season the animals preferred to graze green grass on the ground, but during dry season when there is no blade of grass available, a very common feature in arid region, the animals eat foliage of the trees.

This two tier silvipastoral system is an ideal integrated landuse system that ensures better solar energy capture efficiency and energy flow through the food chain.

Suitable tree species for top feed in silvipastoral system in arid region.

Species	**Crude protein (%)**
Acacia nilotica	13.9
Prosopis cineraria	13.9
Salvadora oleoides	9.6
Zizyphus nummularia	11.7
Acacia senegal	10.3
Albizia lebbeck	20.2
Anogeissus latifolia	7.5
Anogeissus pendula	7.6
Calligonum polygonoides	7.4
Azadirachta indica	12.4
Grewia tenex	10.3
Gymnosporia spinosa	9.3
Prosopis juliflora	21.4
Tecomella undulata	12.2
Leucaena leucocephala	15.2-27.6
Holoptelia integrifolia	13.7

Agrisilvipastoral system

It is a system in which agricultural crops, forest tree species and grasses are grown on same land simultaneously. Indian farming is integrated with animals for carrying out certain farming operations. The common example of agrisilvipastoral system includes intercropping with coconut and horticultural trees if grasses are also grown as intercrops, trees in agricultural fields and trees on farm boundaries. The animal component is variable that is the system of trees in agricultural fields and farm boundaries will have relatively more dependence on animals than the systems of growing agricultural crops with commercial trees.

Hortisilviculture system

It is a system in which horticultural crops particularly fruits crops are grown in an orchard and it is protected by growing tall tree species on its boundary for acting as windbreak, shelterbelts or fillers. Tall growing trees are preferred like *Grewia* spp. *Populus deloides, P. nigra, Salix sp, Casuarina equisetifolia, Syzygium cumini, Tamarindus indica.*

Intercropping of vegetables viz. cowpea, tomato, brinjal, radish, carrot, onion, garlic, cluster bean, ginger, turmeric, etc., in sapota, lime, coconut, mango, guava orchards, is very successful.

Agrihorticulture system

The system in which the farmers continue to take arable crops till trees developed dense canopy or bear fruits and later on it can be converted into horti-pastoral system by introducing suitable improved grass species.

Agrihortisilviculture system

It is a system comprising of a combination of agricultural crops, horticultural crops and forest tree species.

Multipurpose forest tree production (Silviculture)

In this system, the forest tree species are regenerated and managed for their ability to produce not only wood but also leaves and/or fruits that are suitable for fodder and/or food.

Other specialized agroforesty systems are:

Apiculture with trees (Tree + bee keeping)

It is a system under which honeybees are reared on different flowering tree species for production of honey. Honeybees make their hives on the tree branches. They collect honey from flowers. Also, on the ground artificial bee hives are kept for honeybees to collect honey from the necter of the flowers and store inside the hives.

Aquasilviculture (Fish + Tree) or Aquaforestry

It is a system under which silviculture of mangroves and fish is done simultaneously. Tress are also planted on banks of fish breeding ponds.

Agrisilviaquaculture (Tree + Crop + Fish)

In paddy field, fish can easily be reared by planting trees on field bunds or boundary. Altogether, this system is called as agrisilviaquaculture. This system can be followed in high rainfall areas.

To sum up, the agroforestry systems envisage to conserve and improve the site on one hand and to optimise the combined productivity of the forest and the agricultural crops on the other.

Scope of Agroforestry in India

There is tremendous scope of agroforestry in India. A large hectarage is available in the form of farm boundaries, bunds, wastelands where this system can be adopted. This system permits the growing of suitable tree species in the field where most annual crops are growing well. By growing tree species the advantage of soil moisture from deep rooted trees can fulfil their water requirements from the deeper layers of soil. Thus, agroforestry can serve a powerful tool for social forestry in India. Agroforestry has the potential of at least partially meeting the challenge of present scarcities and unemployment, and has a vital role to play in rural development.

Agroforestry can be adopted on agricultural lands, forest lands, marginal and sub-marginal wastelands, lands not presently available for cultivation of arable crops etc.

India has achieved self sufficiency in food production. Now, its attention is becoming more focused on the problems of acute shortage of fodder, fuel and other products. Agroforestry has vast scope in meeting these requirements because it combines with mostly multipurpose tree species which will solve the problems of fuel, fodder, fruits, timber, shade, protection etc. Realizing such scope, an All India Coordinated Research Project on Agroforsetry was initiated in 1983 to initially operate at eight Research Institutes of the Indian Council of Agricultural Research (ICAR) and twelve Agricultural Universities, and now it is being extended to large number of Universities and Institutes.

In fact, population explosion is not at halt resulting in a tremendous increase in the demand for land for raising crops and for housing. Moreover, the forest resources are depleting at a rapid rate as a large population in the vicinity of the forest area depends on it for fuelwood, fodder and other raw materials. The size of land holding is on decreasing order resulting in an increase in number of small and marginal farmers. Not only this, a large number of people are becoming landless and further, the number of landless people are expected to increase in near future. The landless rural people are unemployed. They depend on forest and forest products for their livelihoods.

The most convenient and economic system to produce more from a piece of land is through agroforestry systems (Agrisilviculture, silvipastoral system, agrisilvipastoral system, aquasilviculture, multipurpose tree production etc). These systems are labour intensive hence, employment opportunities to unemployed rural population are bright. This further makes the wide scope for having agroforestry in the rescue of rural poor. In fact, agroforestry assures permanent source of higher income even in extreme adverse conditions. Not only this, inclusion of trees amelioratese edaphic and climatic conditions in favour of crop production.

Since agroforestry systems involve intensive use of land under proper management without deterioration of its fertility that results in more output which adds in national economy. Thus, bright future of agroforestry in India is inevitable.

Present status of Agroforestry in India

Historical background

Agroforestry is as old as the origin of Agriculture. But, the scientific

Approach to this system has been realized recently. Field crops like pearlmillets, legumes and oilseeds have been growing in combination with Khejri (*Prosopis cineraria*) and Bordi (*Zizyphus nummularia*) by the farmers of Rajasthan, Punjab, Haryana, and Gujarat since the time immemorial and similarly, the cultivation of large Cardamon (*Ammomum subulatum*) in combination with *Alnus nepalensis* in Central Himalaya. Various tree crop combinations in Jhum areas of North Eastern Hill region are being practised since long. Tree species viz. *Dalbergia sissoo, Azadirachta indica, Acacia nilotica, Grewia optiva, Morus alba, Ficus* spp. etc. are grown on the borders of fields for meeting demand of timber, fodder, fuel etc. is common practice throughout the country. About 400 running metres boundary is available in a hectare of square piece of land. If this area is further divided into four sub plots it will give 200 running metres more area that is the total area comes to 600 running metres. At one metre distance, 600 plants can easily be planted. After three years of planting, some plants can easily be thinned out for the use as firewood and/or poles. If all plants are felled after 4 or 5 years for poles purpose the income will be very high.

On the same parallel line, Sagwal described the traditional agroforestry as the base for new concept of agroforestry in Western Himalayan region. According to the concept, several tree species like Bhemal (*Grewia optiva*), Kachnar (*Bauhinia variegata*), Khirak (*Celtis autralis*), Ohi (*Albizia chinensis*) and Tuni (*Toona ciliata*), are used to grow on the farm/field boundaries. The species are retained in the field for various purposes. The most important are khair (*Acacia catechu*), Siris (*Albizia* spp.), Drum stick (*Moringa oleifera*), Bahera (*Terminalia bellerica*), Arjun (*T. arjun*), Puna (*Ehretia accuminata*), Ban-oak (*Quercus leucotricophora*), Sal (*Shorea robusta*), Ber (*Zizyphus mauritiana*), Maggar (*Dendrocalamus hemiltonil*), Shisham (*Dalbergia sissoo*), Mulberry (*Morus sp.*) etc. However, no systematic approach was made earlier to identify the systems to improve them for higher productivity, higher economic returns and energy requirements.

Recent developments

A National Seminar on Agro-forestry jointly sponsored by Indian Council of Agricultural Research and Government of Manipur was held at Imphal (Manipur) from 16th to 19th may, 1979. Dr. K.F.S. KING, Director General, International Council for Research in Agro-forestry, Nairobi, delivered the keynote address and suggested the research directions. A good initiation was taken by creating All India co-operating cell in Krishi Bhawan and centers at 20 places in different agro-ecological regions of the country during sixth Five Year Plan.

Agroforestry requires precise research work in a particular area based on climatic and edaphic conditions of the locality. In this direction, though work has already been started but stilt scientific experiments have been lacking so far. However a few research centers have come forward to take up this work. Agri-silviculture research center at Ranchi has initiated comprehensive work under a pilot scheme for research and growing of cash crops in conjuction with main forest species. It is now urgently needed to intensify research efforts not only in the field of agri-silviculture (including farm forestry) but also in the field of other forestry systems, which, in the Indian context, will provide technical base for social forestry programmes of growing trees for fodder, firewood etc. on village common lands and wastelands.

Implications of Agroforestry

The following are main implications of agroforestry.

Agroforsetry requires multidisciplinary approah

Successful adoption of any agroforestry systems need a multidisciplinary approach i.e. it requires knowledge of soil science, agronomy, horticulture, forestry, meteorology, economics etc. In order to evolve the technology of agroforestiy to suit to a particular local conditions, research efforts of the 'Subject/Matter/Specialist' from above mentioned subjects and related one will be needed.

Awareness to local people

While executing agroforestry programme in a particular area, the local people should be made aware about the programme. It has been seen that people when do not get reasonable prices for their produce then they do not like to participate. Lack of facilities discourages them to follow agroforestry systems. There is need to have "Biomass Bank" where cultivators after harvesting and curing biomass can deposit and on need they can have the same. This will prove boon during the period of scarcity where biomass can be utilized with good returns.

Effective extension services

In order to educate cultivators about agroforestry technology different media are needed to provide suitable technology to culivators. Although personal contact with the farmers is the best method for making aware about agroforestry systems and guiding them time to time but this method is rather slow and could cover only limited number of the local people who are participating in it. People should know that they are going to have their basic needs namely fuelwood and fodder at little or no cost. Cultivators should be given knowledge about the selection of suitable tree species preferably multipurpose, and about the awareness of agricultural crops which are compatible. In fact, without participation of local people no plans and programmes can be successfully implimented. When majority of the local people are willing to cooperate in working and protection of the plantation, the programmes will become viable and the great success.

Lack of tree cooperatives

Establishment of tree cooperatives can only enable cultivators to sell their produce on remunerative price. Agroforestry technology may be broadcast on radio, will be much better if it is telecast on T.V. Effective dissemination can also be done by calling cultivators to attend farmers fair, meetings, group discussion that will enable them to learn much better way. Besides, there should be biomass banks in rural areas where villagers can deposit, sell or purchase biomass (wood and foliage). Biomass banks should collect, grade, process and distribute biomass as per need of the local people. Biomass banks will require strong management amd adequate financial back up.

Financial help

Planting of tree species requires money for purchase of saplings, seeds etc. Until and unless sufficient financial help is assured it will become difficult for cultivators to adopt agroforestry system.

Implementation of Government policies backed by adequate financial timely help and participation of local people can only make success in developing wasteland through agroforestry.

Benefits of Agroforestry

The following are main benefits of agroforestry :

- Agroforestry improves environment
  - (a) Nature's gift
  - (b) Ecology and pollution
- Agroforestry is a good source of fuel wood and energy to rural poor.

- Agroforestry provides fodder and feed to animals.
- Agroforestry enriches soil with organic matter and nutrients, and checks soil erosion.
- Agroforestry prevents soil erosion
- Agroforestry provides windbreaks and shelterbelts.
- Agroforestry has employment potential.
- Agroforestry provides raw materials to small cottage industries.
- Agroforestry has better productive potential.
- Agroforestry is a paying occupation.
- Agroforestry creates aesthetic values.

Agroforestry Improves environment

Acute shortage of food, fuel and fodder on one hand and increasing human and livestock population on the other hand have forced the people world over to change their attitude towards exploitation of environment. Recent emphasis on agroforestry at international and national level is one of its consequences.

Nature's gift

Nature has been extremely kind to us in bestowing a large number of birds and useful insect-pests. They, in many cases, are the friends of farmers and an integral part of environment (Table 7.1).

Table 7.1 : Birds and useful insect-pests under agro-forestry system

S.No.	Birds/Insect-pests	Usefulness
1	Great Indian Bustard	These birds are useful in destroying the uncontrolled swarms of locusts.
2	Honey bees	These give honey a food and help to many crops by increasing vigour and vitality of crops.

There are a large number of such birds and insect-pests which help in many ways particularly to the environment.

Plants absorb carbon dioxide and supply oxygen in the process of photosynthesis. At the same time, they reduce pollutants in water and soil. Plants remove gaseous pollutants, dust and dirt particles from the air. Plants are known to have some effects as sound barriers. Most ordinary sound we hear falls in the range of 25 decibels (dBA).

In residential are as, a level of 55 to 60 dBA is desirable during day time and 50 to 57 dBA during the evening hours. The potential value of vegetation as noise abater is deemed very good. A belt of 5 to 6 m wide with a tall, dense trees and shrubs planted close reduces sound 5 to 10 decibels.

Ecology and pollution

The single largest land use in India is agriculture with forestry as next. There is about 43 m hectares of wasteland and a self-generating reinforcing nature built up mechanism of checks and balance among various life forms, through a balance of parasites and predators.

Trees play a vital role in ameliorating the deteriorating environment. They are helpful in reducing noise pollution. They act as dust collectors and air purifiers as they serve as sinks for gaseous pollution, to a degree conserve soil moisture and improve the general environmental conditions. As we are all aware that automobiles exhaust a lot of fumes, carbon dioxide and other harmful gasese into the air, large quantity of toxic industrial effluents added to the rivers and millions of barrels of oil spilled into the oceans, a huge quantity of wood is consumed daily in rural areas for cooking foods, warming houses etc., result in thousand of hectares land cleared. Such an unhappy situation of deforestation causes lot of ecological and social problems. If long term measures for creating better living environmental conditions are not taken up, the environmental hazards cannot be avoided.

Therefore, it is very essential to adopt agroforestry systems where suitable plant species are grown that can add in vegetation cover. Besides, planting of trees on wasteland, barren lands, lands lying vacant along railway track, road, canal, rivers, catchment areas and wherever it is possible to plant trees, should be planted and cared them, till they grow beyond the reach of browsing animals. They are cut only at their most economic stage of growth, and replanting followed without any lapse of time.

Clearing the area by cutting of trees and not following immediately by replanting will lead to ecosystem imbalances. Vegetation maintains these balances.

The following species are helpful in environmental conservation which can be planted according to climatic conditions.

A few tree species for environmental conservation

Acacia auriculiformis A. Cunn
Acacia dealbata
Agathus australis Salish
Ailanthus excelsa Roxb.
Alstonia scholaris Br.
Amherstia nobilis wall.
Anogeisus pendula Edgew.
Araucaria bidwilli Rx.
Araucaria cooki R. Br.
Araucaria cunninghamii Sw.
Araucaria excelsa Br.
Azadirachta indica A. Juss.
Bauhinia acuminata L
Bauhinia alba Buch Ham
Bauhinia corymbosa Roxb.
Bauhinia purpurea L.
Bauhinia tomentosa
Bauhinia triandra Roxb
Bauhinia variegata L.
Bombax ceiba L.
Butea monosperma (Lamk) Taub
Callistemon lanceolatus DC.
Cassia fistula L.
Cassia grandis
Cassia javanica L.
Cassia marginata Roxb.
Cassia nodosa Buch ex ham.
Cassia renigera Wall
Cochlospermum gossypium DC.
Cordial dichotoma
Colvillea racemosa
Delonix regia Raf
Cocous nucifera L.
Erythrina indica lam.
Ficus bengalensis L.
Glyricidia sepium HBK
Hibiscusrosa sinensis
Jacaranda mimosaefolia D. Don
Kigelia pinnata DC.
Lagerstroemia specioesa Pers
Lagerstroemia thorelli Gagnep
Madhuca longifolia Macbride
Mangifera indica L.
Michelia chapaca L.
Milingtonia hortensis L.
Moringa oleifera Lamk
Murraya koenigii K. Spreang
Olea ferruginea Royle
Parkinsonia aculeata L.
Peltophoru ferrugineum Benth
Pinus roxburghi Sar
Pithecellobium dulce Benth
Polyalthia longifolia Thw
Pongamia pinnata Pierre
Populus spp.
Prospis cineraria L.
Pterocarpus indicus Willd
Salix alba L.
Salix tetrasperma Roxb.
Samanea saman Merrill
Santalum album L.
Saraca indica L.
Scheichera oleosa Oken
Sesbania grandiflora Poir
Shorea robusta Gaerth
Spathodia campanulatus Beauv.
Syzygium cumini skeels
Tamarindus indica L.
Tamarix aphylla karst
Tectana grandis L.G.
Terminalia arjuna Bedd
Terminalia belerica Roxb
Terminalia catappa L.
Terminalia chebula Retz
Terminalia paniculata Roth
Tephrosia purpurea soland
Toona ciliato Roen
Zizyphus mauritiana Lamk

A few shrubs for environmental conservation

Acalypha hamiltoniana Hort
Aralia filiciolia Hort.
Caesalpinia pulcherima Sw
Cestrum diurnum
Cestrum nocturnum
Codiaum sp
Cycas circinnalis L.
Dracaena reflexa Lam
Agave spp.
Hibiscus rosa-sinensis
Hibiscus mutabilis
Ixora spp.
Jasminum sambac Ait
Lagerstroemia indica
Nyctanthus arbortristis
Pandanus panderi Hort.
Poinsettia pulcherima Arah
Sansevieria spp.
Tabernaemontana coronaria Willd
Thuja compacta

A few tree species suitable for industrial green-belt

Acacia auriculiformis
Acacia nilotica
Albizia lebbeck
Anthocephalus chinensis
Artocarpus heterophyllus
Alstonia scholaris
Azadirachta indica
Cassia fistula
Cassia siamea
Dalbergia sissoo
Eucalyptus Spp.
Ficus benghalensis
Ficus religiosa
Ficus refusa
Glyricidia sepium
Kigelia pinnata
Mangifera indica
Melia azadirach
Parkinsonia aculeata
Populus spp.
Pongamia pinnata
Polyalthia longifolia
Pithecellobium dulce
Syzygium cumini
Terminalia arjuna
Zizyphus mauritiana

A few shrubs suitable for industrial green-belt

Calotropis gigantia
Helianthus annuum
Dahlia varibilis
Hibiscus rosa sinensis
Hibiscus mutabilis
Jasminum auriculatum
Jasminum grandiflorum
Jasminum sambac
Nerium odorum
Nyctanthus arbortristis
Iopomea palmata
Rosa spp.
Saccharum munja

Agroforestry is a good source of fuelwood and energy to Rural poors

The fuel policy committee estimated fuel demands on all India basis for domestic sector for the year 1978-79 and 1990-91.

Estimated fuel consumption

Fuel	1978-79	1990-91	Estimated consumption in coal replaced in million tonnes 1978-79	1990-91
A. Commercial				
Soft coke	5.0mt	20mt	7.5	30.00
Kerosene	3.5mt	6mt	2.9	49.90
L.P.G.	0.4mt	22mt	3.3	16.61
Electricity	8.0B,Kw	25B.Kw	8.0	25.00
B. Non Commercial				
Fuelwood and wood for charcoal	13mt	122mt	125mt	11.59
Dung cake	65mt	53mt	26.00	21.10
Crop waste	46mt	46mt	43.7	43.70

In order to meet a planned and recorded production demand of 100 million cubic metres of fuelwoocl and assuming a yield of 30 cubic metres per hectare per 10 years, about 3 million hectare per year will be needed to be planted for 10 years. Further, the requirement of fuelwood is estimated to be 225 million cubic metres by 2,000 AD. Shortatge of firewood has been termed a "poorman's energy crisis because it has reduced even intolerably low living conditions of the rural poor.

Unless this problem of fuelwood is solved adequately, alleviation of rural poverty is considered nearly impossible. In this context, the saying that there may not be famine of food but of firewood to cook food seems to be true.

Agroforestry provides Fodder and Feed to Animals

There is an acute shortage of green fodder in many parts. Agroforestry can pave the way to solve the problems as it has great potential. For existing population, as against 100 million tonnes of milk produced annually, the nutritional requirement call for 150 million tonnes and by the end of 2020 A.D., with a projected population of about a billion people, a billion tonnes of milk would be needed. Animal population meeting the milk and meat requirements of growing population will demand increased forage production which can be met out through agroforestry programme.

The following are a few tree species from which leaf fodder and feed can be obtained.

Trees species for Tropical dry areas

Acacia modesta
Acacia nilotica
Ailanthus excelsa
Albizia amara
Albizia lebbeck
Anogeissus pendula
Cordial dichotoma
Cordial rothii
Leucaena leucocephala
Salvadora oleoides
Samanea saman
Tecomella undulate
Zizyphus mauritiana
Zizyphus nummularia

Tree species for Tropical humid areas

Acacia nilotica
Ailanthus excelsa
Albizia amara
Albizia lebbeck
Anogeissus latifolia
Artocarpus heterophyllus
Bauhinia varietgata
Cassia siamea
Cordial dichotoma
Kydia calycina
Moringa oleifera
Morus alba
Terminalia bellerica

Tree species for Subtropical regions

Acacia catechu
Acer oblongum
Aesculus indica
Albizia chinensis
Alnus nepalensis
Alnus nitida
Bauhinia chinensis
Bauhinia purpurea
Celtis australis
Cornus oblique
Ficus racemosa
Grewia optiva
Lasiobema retusum
Robinia pseudacacia
Toona ciliata

Tree species for Temperate regions

Aesculus indica
Alnus nepalensis
Alnus nitida
Celtis australis
Cornus capitata
Fraxinus excels
Grewia species
Lyonia ovalifolia
Olea cuspidate
Pistacia khinjuk
Populus ciliata
Quercus species
Robinia pseudacacia
Salix species

Acacia spp. produce pods, pod bran, seed etc., which are being used as animal feed. *Acacia nitotica* seeds are high in crude protein (18.6%). *Albizia lebbeck* seed, *Leucaena leucocephala* seeds which are at present not used as cattle feed are high in protein (about 30%). They can be utilized in future as sources of cattle feeds. However, subabul seeds contain mimosine which limits their value as animal feed at present. The neem by-products although rich in protein are not commonly utilized as animal feed until debitterization process develops. Mahua trees yield two products besides leaves, *viz.*, mahua flowers and mahua seed cake. Both these products can be used as animal feed. In rubber growing areas, rubber seed meal is available after oil extraction for feeding animals. *Prosopis juliflora* pods can be collected from Gujarat. It is a potential new source of cattle feed.

The following fruit species may be planted according to climatic conditions. These species not only provide fruits but also many of them provide leaf fodder during scarcity. They provide adequate firewood from their prunings.

Fruit species for Tropical dry regions

Aegle marmelos
Capparis aphylla
Carissa carandas
Ceratonia siliqua
Cordial dichotoma
Cordial rothi
Emblica officinalis
Manikara hexandra
Moringa oleifera
Pithecellobium dulce
Salvadora oleoides
Tamarindus indica
Zizyphus mauritiana

Fruit species for Tropical humid regions

Anacardium occidentale
Areca catechu
Mangifera indica
Manilkara hexandra

Artocarpus heterophyllus
Carissa congesta
Cocos nucifera
Cordial dichotoma
Emblica officinalis
Feizoa sellowiana
Madhuca longifolia
Moringa oleifera
Morus alba
Parkia roxburghi
Pithecellobium dulce
Syzygium cumini
Syzygium jambos/jambolana
Tamarindus indica

Fruit species for Sub-tropical regions

Castania sativa
Carylus colurna
Emblica officinalis
Litchi chinensis
Mangifera indica
Prunus armeniaca
Prunus communis
Prunus persica
Psidium guajava
Pyrus communis

Fruit species for Temperate regions

Berberis
Berberis petioleris
Juglans regia
Olea cuspidate
Prunus armeniaca
Prunus cerasus
Prunus communis
Prunus persica
Punica granatum
Pyrus communis
Pyrus malus
Rhus semialata
Rubus species

Agroforestry enriches Soil with Organic matter and Nutrients

There is a great symbiosis between agriculture and forestry. Agroforestry can play an important role to meet growing demand of organic manure. The unproductive and degraded land can meet out substantial requirement in the form of leaf litter instead of cow-dung manure, which is in shortage as the same is being used for both fuel and manuring purposes.

Annual leaf litter production (kg/ha/yr)

Tree Species	Mean Values 1965-66 to 1970-71
Babul (*Acacia nilotica*)	8056
Bent Bamboo (*Dendrocalamus sp.*)	6033
Siris (*Albizia sp.*)	4994
Teak (*Tectona grandis*)	4631
Eucalypts (*Eucalyptus spp.*)	1764

Annual leaf litter production is highest in *Acacia nilotica* (Babul) whereas lowest in *Eucalyptus* sp.

Economics of forest leaf litter

Various	Tree Species				
Components	**Babul**	**Bamboo**	**Siris**	**Teak**	**Eucalyptus**
Leaf litter					
Produced (kg/ha)	8000	6000	5000	4600	1800
N-content (kg/ha)	64	48	40	35	14
P_2O_5 (kg/ha)	48	36	30	28	10
K_2O (kg/ha)	80	60	50	48	18
Value of N_2 (Rs)	192	144	120	105	43
Value of P_2O_5 (Rs)	144	108	90	84	32
Value of K_2O (Rs)	80	60	50	46	18
Total Value of (N_2+P_2O_5+K_2O) Rs	416	312	260	235	93

i) Contents of NPK in dry compost has been taken as 0.8, 0.6 and 1.0 per cent, respectively.

ii) Value of NPK has been taken as Rs 3/-, 3/- and 1/- per kg, respectively.

Comparative value of chemical fertilizer and dry litter compost

Species	Dry litter produced kg/ha/yr	Cost of equivalent chemical fertilizer Rs.	Cost of dry litter Rs.
Babul	8000	416	320
Bamboo	6000	312	240
Siris	5000	260	200
Teak	4600	235	185
Eucalyptus	1800	94	74

Note: Cost of dry compost has been taken at Rs 4/100 kg as per prevailing local rates.

In the use of inorganic fertilizer, raw material is not renewable. Biogas plant yields both fuel and richer organic manure.

A large number of legume plant species like *Acacia spp., Leucaena* spp., Groundnut, Peas, Beans, Lentil, Alfalfa, *Lupinus* spp., and a few non-legume species like *Alnus cripsa, A. glutinosa, A. incana, A. rubra* and *A. rugosa, Cariaria* spp., *Casuarina* spp., *Ceranothus* spp., *Elaeagnus* spp., *Myrica* spp., are known to synthesize considerable quantity of nitrogen from the atmosphere.

Agroforestry prevents Soil erosion

Growing of trees and shrubs with agricultural crops will help in preventing soil erosion. (i.e. by wind and rainwater). Agricultural crops are short duration whereas tree species are long-lived and, therefore, in the presence as well as in the absence of annual crops, they act as better agents for soil conservation. In fad, each tree acting as a miniature dam, reduces the velocity of winds and rushing rain water. The canopy of the tree is able to reduce raindrop impact, therefore, protecting the soil from splash erosion.

The ground litter is also known to enhance water percolation into the ground and hence reduce run-off and rill erosion. It is also expected that roots of trees planted along the terrace edges and embankments would bind soil together, which would enhance terrace stability.

The following species when planted at high density check soil erosion up to a great extent.

Tropical Dry Regions

Acacia catechu
Acacia nilotica
Acacia Senegal
Acacia tortilis
Ailanthus excelsa
Albizia lebbeck
Azadirachta indica
Cassia auriculata
Cassia siamea
Dendrocalamus strictus
Lowsonia inermis
Leucaena leucocephala
Pithecellobium dulce
Prosopis cineraria
Prosopis juliflora
Tamarix troupii
Tephrosia purpurea

Tropical Humid Regions

Acacia auriculiformis
Acacia catechu
Acacia nilotica
Ailanthus excelsa
Erythrina species
Eucalyptus species
Glyricidia sepium
Gmelina arborea

Albezia amara
Albizia lebbeck
Anacardium occidentale
Cassia siamea
Dalbergia sissoo
Dendrocalamus strictus
Pandanus odoratissimus
Pitthecellobium dulce
Pangamia pinnata
Salix tetrasperma
Tephrosia purpurea

Subtropical Regions

Alnus nepalensis
Alnus nitida
Bombax ceiba
Cariaria nepalensis
Celtis australis
Cupressus cashmeriana
Cupressus torulosa
Dendrocalamus strictus
Erythrina superb
Eucalyptus species
Exbrucklandia populanea
Fraxinus species.
Launea coromendelica.
Macaranga indica
Populus deltoids.
Populus gamblei.
Populus nigra.
Robinia pseudacacia.
Salix species.
Sapium insigna.
Stenolobium stans.
Woodforadia fruticosa

Temperate Regions

Alnus nepalensis
Alnus nitida
Arundinaria fulcata
Buddleia asiatica
Cariaria nepalensis
Celtis australis
Cotoneaster baciliaris
Cupressus cashmiriana
Cupressus torulosa
Deutzia corymbosa
Fraxinus species
Indigofera species
Populus ciliata
Populus yunnansis
Robinia pseudacacia
Spiraea canescens
Spiraea corymbosa

Agroforestry provides Windbreak and Shelterbelt

Windbreak is a single or double rows of quick growing tall tree species or a combination of tall and semi-tall species planted at high density either round the boundary or on the ward-wind direction of the farm or orchard in order to provide effective protection to the main crop against adverse effects of high velocity of hot and cold winds.

Developed countries like USA and USSR have demonstrated that agriculture production can be increased to the extent of 20% by using windbreak through agroforestry.

Shelterbelt usually consists of several rows of trees and shrubs. There are mainly three types of shelterbelt such as;

- *Wind-proof* it is a dense rows of trees and shurbs.
- *Permeable*: shelterbelt which allows air flow.
- *Porus*: shelterbelt which is having some gaps in planting.

Among three types of shelterbelts, permeable type is reported to be the best as it allows air flow to the crop plants. The effectivness of shelter- belt depends how fast species are growing and their life span. The thickness of shelterbelts depends on the general wind velocity of the locality. That is the thickness or number of rows increases with increase in wind velocity. Usually, shelterbelts occupy the area of about 1.5 to 3.5 per cent depending upon climatic conditions. Planting distance of 3×2 m (row × plant) will suffice the purpose. Wide space is kept in dry climatic conditions as compared to humid climatic conditions.

In India, one study on shelterbelt plantation has shown that yield per unit of land for groundnut, redgram, and bajra (pearlmillet) under shelterbelt conditions were higher by 75.48 and 63%, respectively as compared to control. In Rajasthan, *Prosopis cineraria* is being grown in agricultural fields which besides supplying much needed fuel and fodder, also checks soil erosion and improves agricultural crop yield. The windbreaks reduce the wind velocity, decrease evaporation, transpiration and increase moisture conservation resulting in more production.

Characteristics of tree species for shelterbelts

The following are some of the important characteristics of tree species suitable for planting as shelterbelts.

- Tree species should be fast growing, tall, of uniform shape with good number of branches and dense foliage.
- Tree species shound be hardy against drought, frost, extreme temperatures, windthrow and snow breakage.

- Tree species should have ability to develop strong and deep root systems so that they do not compete for moisture and nutrients with agricultural crops.
- Tree species should be comparatively long-lived, require minimum care and maintenance.
- Tree species should have high survival percentage and easy to establish upon transplanting.
- Tree species should be useful as firewood, small timber and leaf fodder.
- Tree species for shelterbelts should not be self-sown.
- Tree species should be evergreen or partial deciduous.
- Tree species should not be susceptible to insect-pests and diseases. Otherwise they will act as host plants.

Care of windbreak and shelterbelts

- While planting care should be taken that the position of the plants (in case of double row or more) in the succeeding row should be alternate to the plants of preceeding row. This type of planting will be more effective in creating barrier for wind velocity.
- Usually, the effect of windbreak or shelterbelt is about four times of the plant height.
- Trees of shelterbelt should be protected from browsing animals.
- They are also protected from pests and diseases.
- The growth rate of the trees can be enhanced by intercultivating, removing the weeds and by applying the manures and irrigation.
- Sometimes, there is need to dig a ditch one metre deep parallel to the belt at a distance of 2 m from the row of agricultural crops. The ditch may be refilled and again open in next year, and repeated like this. This important operation restricts horizontal development of roots towards agricultural crops due to root pruning, and thus, avoids competition for nutrients and soil moisture.
- In the first year of planting, application of some water is essential for cent per cent survival and early start of growth. The quantity and frequency of watering depends on climatic conditions, soil types, tree species etc.

Demerits of windbreak and shelterbelts

- They occupy the land which could have been under the main crop.
- They impart shading effects and many times interfere with the agricultural operations.
- When trench is not made, their roots may compete with near by plants of the main crop for soil moisture and nutrients.

Tree suitable for windbreaks

For dry and arid regions

Acacia auriculiformis
Ailanthus excelsa
Albizia lebbeck
Anogeissus latifolia
Azadirachta indica
Casuarina equisetifolia
Delbergia sissoo
Eucalyptus citriodora
Eucalyptus hybrid
Moringa oleifera
Peltophorum ferrugineum
Polyalthia longifolia
Pongamia pinnata
Prosopis juliflora
Sesbania grandiflora
Syzygium cumini
Tamarindus indica
Tamarix articulata

For coastal areas

Tree species

Acacia auriculiformis
Ailanthus malabarica
Anacardium occidentale
Calophylum inophyllum
Cassia inophyllum
Casuarina equisetifolia
Grevillea robusta
Mangifera indica
Mesua ferrea
Polyalthia longifolia
Pongamia pinnata
Tephrosia purpurea

Shrub species

Agave americana
Cassia auriculata
Dodonia viscosa
Euphorbia terucalli
Glyricidia sepium
Jatropha spp.
Sesbania aculeate
Sesbania aegyptica
Thevetia neirifolia
Vitex negundo

For temperate regions

Acacia dealbata
Acacia decurrens
Acacia melanoxylon
Acer oblongum
Albizia lebbeck
Alnus nitida
Bombax ceiba
Carya illionensis
Castenea sativa
Celtis australis
Çupressus macrocarpa
Eucalyptus globules
Eucalyptus citriodora
Franela rhomboidea
Grevillea robusta
Grevia otiva
Hakea salinga
Juglans regia
Melia azedarach
Populus spp.
Prunus cornuta
Robinia pseudacacia
Sapindus mukorosii
Terminalia arjuna
Ulmus laevigata

Species suitable for shelterbelts

Acacia catechu
Acacia leucophloea
Acacia modesta
Acacia nilotica
Agave species
Ailanthus excels
Albizia lebbeck
Azadirachta indica
Cassia viscose
Cassia viscose
Dodonaea viscose
Euphorbia terucalli
Glyricidia sepium
Jatropha curcus
Leucaena leucocephala
Melia azedarach
Opuntia dillenti
Peltophorum ferrugineum
Pongamia pinnata
Prosopis juliflora
Salvadora oleoides
Sesbania aculeata
Sesbania aegyptica
Syzygium cumini
Tamarix articulata
Thevetia neriflolia
Vitex negundo

For coastal areas

Ailanthus malabarica
Albizia lebbeck
Alstonia scholaris
Amherstia nobilis
Anacardium occidentale
Azadirachta indica
Glyricidia sepium
Lagerstroemia indica
Lagerstroemia spesiosa
Lagerstroemia thorelli
Mangifera indica
Pongamia pinnata

Bauhinia purpurea
Bauhinia variegata
Brownea grandiceps
Calophyllum inophyllum
Cananga nodosa
Casuarinas equisetifolia
Colvillea racemosa
Coruouptia guianensis
Erythrina cristagalli
Pterospermum acerifolium
Putranjiva roxburghii
Samanea saman
Saraca indica
Spathodea companulata
Stenocarpus sinuatus
Sterulia balanghas
Terminalia catappa

For temperate hill areas

Acacia melanoxylon
Alnus nitida
Cuppressus macrocarpa
Cuppressus torulosa
Dodonaea viscosa
Eucalyptus globulus
Franela rhomboidea
Grevillea robusta
Hakea salinga
Robinia pseudacacia
Ulmus laevigata

Agro-Forestry has Employment Potential

There will be need to view agroforestry and social forestry in the light of essential requirements of much increased employment to the people. There were 504 employment exchanges upto 1982 in the country which gave opportunity to persons seeking employment which is at a rising rates. The figure in 1982 exceeded 17 millions, of whom about 50 per cent were educated. But the total number of unemployed persons is at least 5 times more than this number. About 70% people subsist on land. It is unlikely that the number of people living on land is going to be reduced. Agroforestry provides larger means of providing a higher level of employment particularly in rural areas. Because agroforestry is having a number of labour intensive activities. For instance, in energy plantation programme a number of activities are carried out such as nursery development, digging of pits and trenches, sowing of seeds, watering, weeding, transplanting, aftercare, harvesting, transportation, handling and processing. It has been estimated that planting on one hectare of land generates employment of about 300 man days. Establishment of a change of biomass based industries will generate employment both for the skilled and non-skilled and thus prevent outflow of income from village to cities.

Agroforestry Provides Raw Materials for Small Cottage Industries

At present, minor forest products form about 60% of the total annual export of forest products. Imports of fatty oils have increased tremendously. Raising of oil seed yielding tree species and essential oils bearing especially aromatic plants and grasses under agro-forestry programme will go a long way in not only augmenting the indigenous production of fatty and essential oils but will also open venue for establishment of small scale units. Some of the fibres are being consumed by local industries and are also being exported earning Rs. 4.75 crores annually, and similarly, India is importing tannin extract worth Rs. 10 crores annually. Preferential selection of tannin yielding species under agro-forestry will create favourable conditions for setting up of small scale units thereby fetching better return.

In India, present production of pulp, paper and board is about 1.5 million tonnes per year. It is envisaged that demand for pulp, paper and board will go up to three times by next coming years. The increased need of pulp wood, could be met with by agroforestry and social forestry through raising fast growing tree species alongwith agricultural crops. Trees like *Eucalyptus*, Poplars and Subabul are acceptable pulp for paper making.

Species for Paper Industry

Acacia auriculiformis
Acacia nilotica
Acer campbellic
Aegle marmelos
Ailanthus excels
Albizia lebbeck
Alnus nepalensis
Anthocephalus chinensis
Bambusa aurandinacea
Bombax ceiba
Broussonetia papyrifera
Boswellia serrata
Butea monosperma
Canarium euphyllum
Casuarinas equisetifolia
Eucalyptus hybrid
Exbucklandia populnea
Gmelina arborea
Hevea brasiliensis
Kydia calycina
Lagerstroemia speciosa
Launea coromandelica
Leucaena leucocephala
Madhuca longifolia
Moringa oleifera
Morus alba
Populus ciliata
Populus deltoids
Querus floribunda
Sesbania grandiflora

Chloroxylon swietenia
Dalbergia sissoo
Eucalyptus globulus
Eucalyptus grandis
Shorea robusta
Sterulia urens
Terminalia myriocarpa
Trewia nudiflora

Species for Tanning Industry

The following species of which bark is used for obtaining tannins.

Acacia decurrens
Acacia leucophloea
Aleurites triloba
Artocarpus heterophyllus
Avicennia officinalis
Bauhinia malabarica
Bauhinia purpurea
Bauhinia vareigata
Bridelia retusa
Calophyllum inophyllum
Cassia siamea
Casuarinas equisetifolia
Rhus mysurensis
Psidium guajava
Shorea robusta
Tamarindus indica
Terminalia arjuna
Terminalia catappa
Tona ciliate
Woodfordia fruticosa

Species suitable for Silk Industry

The following are host tree species for silk worm rearing.

Anogeissus latifolia
Dalbergia sissoo
Ficus tsjakela
Lagerstroemia speciosa
Machilus bombycina
Machilus odoratissima
Madhuca indica
Madhuca longifolia
Michelia campaca
Morus alba
Ricinus communis
Shorea robusta
Terminalia arjuna
Terminalia tomentosa
Tetranthera monopetala
Zizyphus mauritiana

Species suitable as a host for Lac production

Acacia catechu
Acacia nilotica
Butea monosperma
Schleichera oleosa
Shorea robusta
Shorea talura
Zizyphus mauritiana

Species suitable for basket making

Acacia nilotica
Arundinaria falcata
Bambusa arundinacea
Dendrocalamus strictus
Salix alba
Salix babylonica
Salix tetrasperma
Tamarix articulata
Tamarix dioica

Species suitable for match Industry

Ailanthus excelsa
Alstonia scholaris
Anthocephalus indicus
Bombax ceiba
Endospermum malacensa
Garuga pinnata
Machilus macrantha
Mangifera indica
Ficus lambertiana
Polyalthia longifolia
Populus tremula
Sterculia campanulata
Tetramelea nudiflora
Tilia saponica
Trewia nudiflora

Species suitable for pencil making Industry

Alnus nepalensis
Alstonia scholaris
Cuppressus torulosa
Juniperus virginiana
Pinus wallichiana
Salix tetrasperma
Sapiun sebiferum

Species suitable for furniture and cabinet making

Adina cordifolia
Albizia lebbeck
Artocarpus chaplasha
Bridelia retusa
Chloroxylon swietenia
Dalbergia latifolia
Dalbergia sissoo
Diospyros melanoxylon
Dipterocarpus sp.
Gmelina arborea
Heritiera minor
Madhuca indica
Madhuca longifolia
Mangifera indica
Melia azedirach
Michelia champaca
Ougenia dalbergioides
Pterocarpus acerifolium
Pterocarpus dalbergioides
Pterocarpus marsupium
Pterocarpus acerifolium
Shorea robusta
Soymida febrifuga
Tectona grandis
Terminalia alata
Terminalia procera

Species suitable for Wood carving

Albizia lebbeck
Crataeva nurvala
Dalbergia latifolia
Dalbergia sissoo
Diospyros melanoxylon
Gmelina arborea
Pterocarpus indicus
Pterocarpus santalinus
Santalum album
Tectona grandis
Wrightia tomentosa

Species suitable for Plywood making

Adina cordifolia
Ailanthus species
Albizia lebbeck
Alstonia scholaris
Anthocephalus indicus
Bombax ceiba
Boswellia serrata
Dalbergia latifolia
Dalbergia sissoo
Evodia roxburghiana
Gmelian arborea
Grevilea robusta
Holoptelia integrifolia
Juglans regia
Launea grandis
Lagestroemia lanceolata
Mangifera indica
Michelia champaca
Pterocarpus dalbergioides
Pterocarpus marsupium
Shorea assamica
Shorea robusta
Swietenia mahagoni
Syzygium cumuni
Tectona grandis
Terminalia alata
Terminalia bellerica
Toona ciliata
Vateria indica

Agroforestry has better Productive potential

Most of the forests are distributed in traditionally refractory sites. Fertility position of agricultural land is comparatively diverting animal dung from present major use as source of cooking or heating energy. The comparison between wood production by traditional forests and agroforestry has been given in the following table.

Wood production by different sources

Sr. No.	Year	Wood production by traditional forests million m³ (r)	Wood production by agroforestry million m³ (r)
1.	1983	200	30
2.	1985	210	45
3.	1990	160	100
4.	1995	105	160
5.	2000	103	200

(r) Round wood

As indicated in Table 6 by 1995 agroforesfty must be capable of producing 160 million cubic metres of round wood i.e. 155 million ha of agricultural land is to produce one cubic metre of wood/hectare which on the fact of programme is not difficult achievement.

Agroforestry is a paying occupation

The economics of some planting models of a few trees and grass species was workedout on the basis of per hectare per year. A few examples are shown in the Table-7.

Economics of few species

Trees species	Income (Rs)		
	Single row planting	Multiple row planting	Close planting
Poplar (*Populus* sp.)	1, 286*	5, 144*	5, 258
Robinis (*Robinia pseudacacia*)	3, 518*	12, 987*	17, 475
Bamboo (*Dendrocalamus* sp.)	3, 780*	3, 600*	4, 400

* Income in addition to the agricultural crops.

The following para indicates how agroforestry benefits mutually. The legume species shed huge quantity of leaves and also synthesize elemental nitrogen in their root nodules which is available to agricultural crops and soil, Poplar plantations (*Populus deltoides*) in the Tarai area of Uttar Pradesh have proved most ideal for successful agri-silviculture operations and intercropping of poplar plantations with paddy in the rainy season and wheat in the winter season. Poplar is a deciduous tree which sheds leaves during wheat growing season. Suitable size is also maintained through pruning which does not interfere with solar radiation. In Jammu and Kashmir, legume fodder, maize and grasses are successfully raised as intercrops in poplar plantation. Poplar plants are benefited by cultural operations, irrigations etc. done in intercrops. Another excellent examole of mutual benefits has been quoted with *Acacia albida* (Africa savanas tree) it sheds its leaves in the onset of rains and retains leaves during dry season. Both the leaves and pods are used as fodder for cattle. Not only this, cattle take shelter underneath *Acacia albida* when all other trees are leafless. The dense leafy crown prevents wind erosion. The cattle dung leaves enrich the soil.

Agroforestry Creates Aesthetic Value

Agroforestry system creates aesthetic values as it includes planting of various multipurpose trees and shrubs through their green foliage, varying colours and shades of flowers, fruits, which also emit fragrance. Tress also help to screen out something unsightly from view e.g., parking lotful of automobiles. For this plants should be tall enough to block the lines of sight of the viewers to the objectionable elements. Also, plants can effectively be used as a structural element in a landscape. Some plants can be used as a curtain to beautify the walls, for instance veronia, Parda bell (*Voronia elaeagnifolia Cav*). A few trees and shrubs suitable for different purposes and situations are given as under:

A few important shade and foliage trees

Acacia auriculformis
Albizia lebbeck
Alstonia scholaris
Araucaria cookie
Averrhoa carambola
Azadirachta indica
Calophyllum inophyllum
Cassia javanica
Cassia nodosa
Cassia renigera
Cassia siamea
Casuarinas equisetifolia
Ceiba pentandra
Dalbergia sissoo
Ficus benjamina
Ficus krishnae
Grevillea robusta
Jacaranada mimosaefolia
Kigelia pinnata
Melia azedarach
Millettia ovalifolia
Mimusops elengi
Polyalthia longifolia
Pongamia pinnata
Pterocarpus dalbergioides
Putranjiva roxburghii
Saraca indica
Tamarindus indica

Emblica officinalis
Eucalyptus spp.
Ficus benghalensis
Ficus infectoria
Tectona grandis
Terminalia arjuna
Terminalia catappa

Suitable tree species which can be mixed for aesthetic effect for the plains

Acacia auriculiformis
Azadirachta indica
Dalbergia sissoo
Madhuca indica
Mangifera indica
Mesua ferrea
Pongamia pinnata
Putranjiva roxburghii
Syzygium cumini
Tamarindus indica

For the hills

Aesculus indica
Celtis australis
Cupressus species
Grevillea robusta
Juglans regia
Pinus deodara
Pinus greggii
Populus species
Salix species

Quick growing trees

Acacia auriculiformis
Acacia decurrens
Adenanthera pavoni
Anthocephalus indicus
Bignonina megapotamica
Bombax ceiba
Cassia fistula
Casuarinas equisetifolia
Cordial sebestina
Delonix regia
Eucalyptus citriodora
Kigelia pinnata
Kydia calycina
Leucaena leucocephala
Millingtonia hortensis
Muntingia hortensis
Samanea saman
Sesbania grandiflora
Sterculia alata
Sterculia villosa
Solanum macranthum
Tabebuia rosea
Tabebuia spectabilis

Trees suitable for flower purposes

Yellow flowered trees

Acacia auriculiformis
Acacia baileyana
Erythrina parcelli
Laburnum anagyroides

Acacia dealbata
Acacia decurrens
Acacia implexa
Anthocephalus indicus
Bauhinia tomentosa
Cassia calliantha
Cassia fistula
Michelia campaca
Parkinsonia aculeata
Peltophorum ferrugineum
Erythrina indica
Schizolobium excelsum
Tabebuia spectalilis
Thespesia populnea

Red flowered trees

Bombax ceiba
Brownea grandiceps
Erythrina blakei
Erythrina laurifolia
Erythrina indica
Saraca indica
Spethodea campanulata
Wrightia coccinea

Scarlet flowered trees

Barringtonia acutangula
Brassia actinophylla
Brownea coccinea
Butea monosperma
Callistemon lanceolatus
Delonix regia
Stenocarpus sinuatus
Sterculia acerifolia

Pink flowered trees

Bauhinia triandra
Cassia grandis
Cassia javanica
Cassia nodusa
Cassia renigera
Hibiscus collins
Kleinhovia hospital
Lagertroemia speciosa
Samanea saman
Sesbania grandiflora

Blue flowered trees

Bolusanthis speciosus
Jacaranda mimosaefolia
Solanum grandiflorum
Solanum macranthum

White flowered trees

Albizia lebbeck
Bauhinia acuminata
Calophyllum inophyllum
Kydia calycina
Madhuca indica
Magnolia grandiflora
Magnolia pterocarpa
Mesua ferrea
Millingtonia hortensis
Mimusops elengi
Moringa oleifera
Oncoba spinosa
Plumeria alba

List of few trees which usually have peak flowering in a particular season

Spring season

Acacia baileyana
Bauhinia acuminate
Bombax ceiba
Brownea coccinea
Butea monospenma
Callistemon lanceolatus
Erythrina indica
Michelia champaca
Moringa oleifera
Plumeria alba
Saraca indica
Spathodea campenulata
Sterulia acerifolia

Summer season

Acacia decurren
Acacia salinga
Albizia lebbeck
Barringtonia acutangula
Bauhinia acuminata
Brownea grandiceps
Calophyllum inopyhyllum
Cassia fistula
Cassia javanica
Cassia nodosa
Cassia renigera
Delonix regia
Erythrina parcellii
Jacaranda mimosaefolia
Laburnum anagyroides
Lagerstroemia speciosa
Magnolia grandiflora
Mesua ferrea
Mimusops elengi
Oncoba spinosa
Peltophorum ferrugineum
Pterocarpus dalbergioides
Stenocarpus sinuatus
Tabebuia spectabilis
Wrightia coccinea

Rainy season

Acacia auriculiformis
Acacia implexa
Bauhinia tomentosa
Cassia calliantha
Lagerstroemia thorelli
Millingtonia hortensis
Peltophorum ferrugineum
Samanea saman

Winter season

Bauhinia triandra
Callistemon lanceolatus
Kydia calycina
Sesbania grandiflora

Year round

Solanum grandiflorum
Solanum macranthum
Thespesia populnea

Trees suitable for sacred places

Aegle marmelos
Amherstia nobilis
Araucaria cookie
Callistemon lanceolatus
Ceiba pentandra
Crataeva nurvala
Diospyros peregrina
Gmelina arborea
Samanea saman

Agro-forestry has the potential of atleast partially meeting these challenges of present scarcities, losses and unemployment as it has a vital role to play in the rural development. Package of technology has to be developed for adoption in various agro- climatic zones after taking into account the needs, customs and traditions of local population. Training to ultimate users of technology should be tailored around the broad guidelines. Seeing is believing, Believing testing, Testing for friability, Trying for feasibility and economic viability. Training and visit system should also be adopted for popularising the concept of agro-forestry among the farmers and foresters. Being varied nature of climatic conditions, agro- forestry has a good scope in India. There is an additional advantage of forest produce of agro-forestry that if, a farmer does not get the expected price then he need not to harvest the tree crops and can retain for longer period as the commodity is not perishable and tree crops can be harvested when ready irrespective of prevailing prices in the market. Therefore, agro-forestry can serve better the posterity in a sound manner which will be true service to the nation.

❑❑❑

Chapter-8

Integrated Farming Systems and Profitability

Crop livestock integrated farming systems as well as agro processing is expected to increase productivity, efficiency of small farmers and would create multiple livelihood opportunities.

National policy for farmers, 2007

- Appropriate development concepts have to be elaborated and adjusted to the local conditions to meet the particular requirements of the farmers
- Integrated Farming system can be regarded as a promising option to create additional cash-income and thus alleviating poverty, by an increase in productivity through a multiple use of farmland
- The IFS approach is holistic, multidisciplinary, problem solving, location specific and farmer-oriented, and it treats agriculture as a complex interaction between natural and social phenomena
- Integrated farming systems (IFS) approach serves as a good model for helping subsistence farmers achieve a reliable food supply, increased income, and greater environmental sustainability.

Farming system-Definitions

- The household, it's resources, resource flows and their interaction at the individual levels are together referred to as a farming system. FAO, 2001
- Farming system refers to particular arrangement of farm enterprises that are managed in relation to physical, biological

and socio-economic environment keeping in view and in accordance with the farmers' goals, preferences and resources.

Shanor *et al.*, 1982

- Farming system is a complex inter-related matrix of soil, plant, animals, implements, power, labour and other inputs controlled in part by farm families at centre and influenced to varying degrees by political, economic, institutional and social forces that operated at many places. Mahapatra, 1992

What is Integrated Farming System (IFS)

Farming system which integrates natural resources and regulation into farming activities to achieve maximum replacement of off-farm inputs, secures sustainable production of high quality food and other products through ecologically preferredtechnologies, sustained farm income, reduction of sources of present environment pollution generated by agriculture and sustains the multiple function of agriculture

When different enterprises are dependent, complementary and supplementary to each other, they interact among themselves and affect the others. Such a mixed farming system is termed an "integrated farming system".

Aims we can achieve

- Efficient recycling of farm and animal waste.
- Minimize the nutrient losses.
- Maximize of the nutrient use efficiency.
- Following of efficient cropping systems and crop rotations.
- Complementary combination of farm enterprises.
- Increase the productivity and sustainability of the system.
- Increase the profitability of the farm.

Unsustainable system

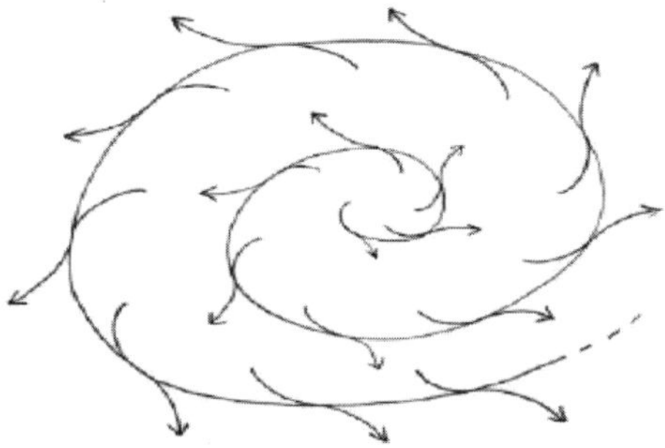

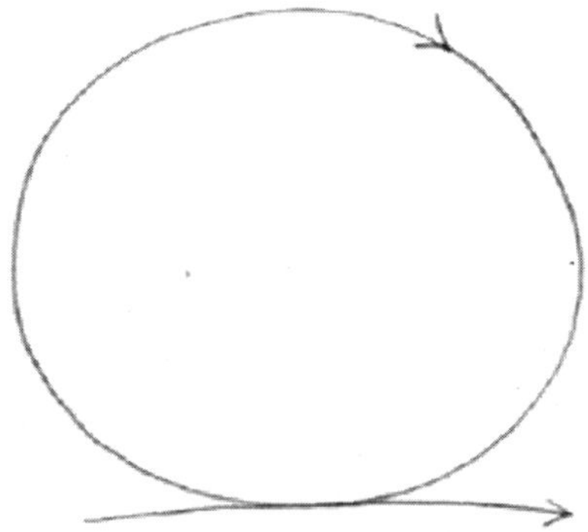

Swallows up the earth's resources and exports massive amounts of wastes and entropy Sustainable system

The sustainable system closes the energy and resource use cycle, maximizing storage and internal input and minimizing waste, rather like the life cycle of an organism that is autonomous and self-sufficient.

Elementary Integrated Farming system

Livestock is incorporated to close the circle, thereby minimizing external input, while maximizing productivity and minimizing wastes exported to the environment.

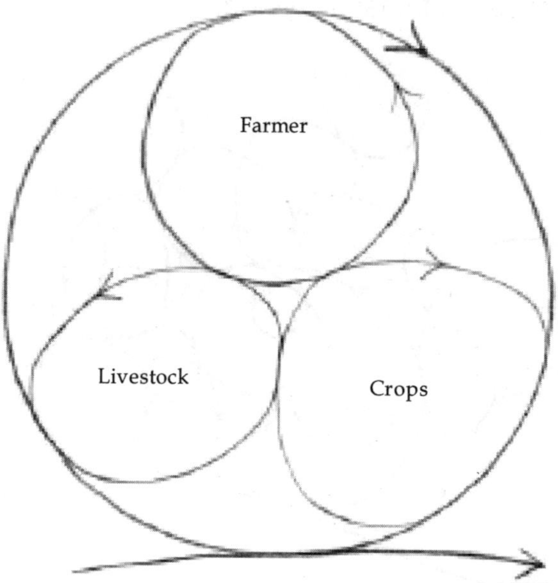

Indigenous integrated farming system

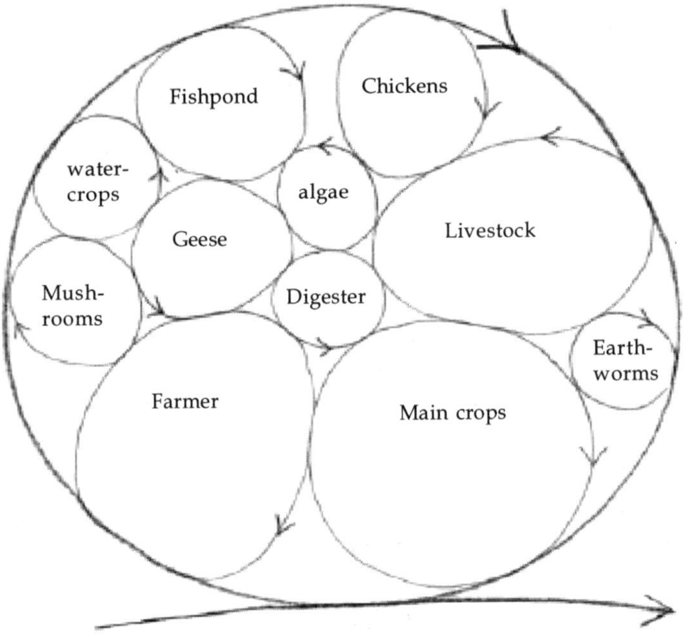

Increasing productivity by incorporating more lifecycles into the system.

Highly productive sustainable system

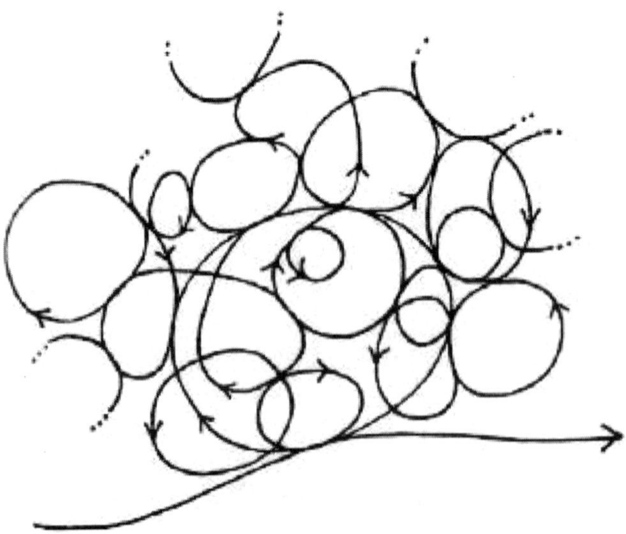

Possible complementary enterprises

- Crop production
- Fruit production
- Agro-forestry
- Vegetable production
- Ornamental flower
- Seed production
- Aromatic and medicinal plant
- Post harvest processing
- Biogas production
- Sericulture
- Dairy
- Goat rearing
- Sheep rearing
- Poultry
- Fish rearing
- Rabbit farming
- Apiaries
- Duck farming
- Quail production
- Mushroom production
- Piggery

Resource recycling on on-station farming system unit, OUAT, Bhubaneshwar

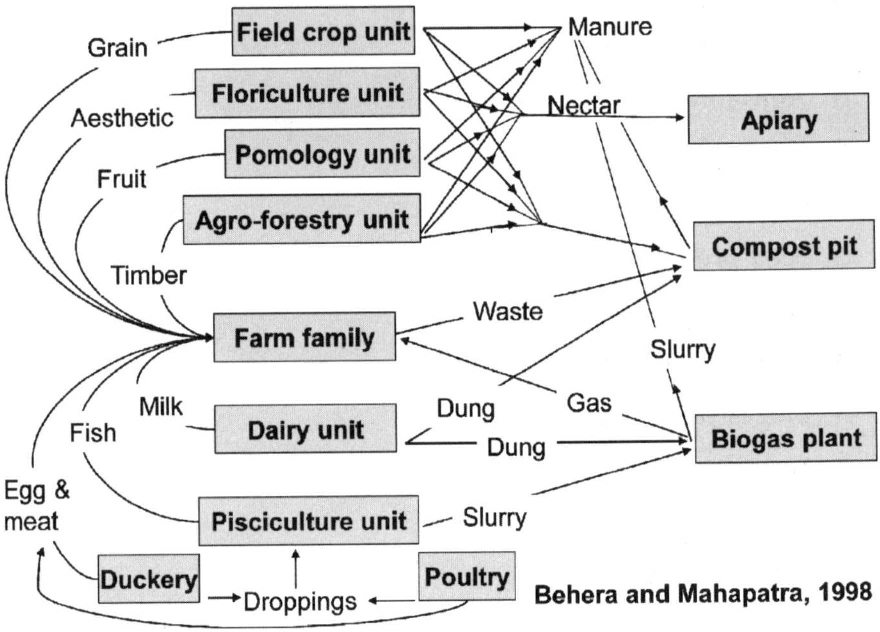

Behera and Mahapatra, 1998

Cash flow

- 1st year, Net returns- fixed capital (40,340-25,680) 14,660/-
- 2nd year onwards (Rs.1500 from banana) 41,840/-
- 5th year onwards *(including income from coconut, banana and others)* 47,840/-

Economics and benefits

- Can produce 12 to 18t of food crops, 0.3 to0 .4t of fish and 3 to 4t of rice straw that can be used as fodder annually which is 15 to 18 times more than that of conventional farming.
- Net income in the 1st year is Rs. 15,000 excluding home consumption. The income doubles from 3rd year. In the long term the expected income from timber and other trees will be about Rs. 500,000.
- Caters to balanced dietary need of farm family.
- Generates additional employment of 450 to 500 mandays/year, thereby restricting the migration of farming community.

Advantages

- Increases in productivity and sustainability of system.
- Increases in profitability of farm.
- Provides balanced food.
- Reduces environmental pollution.
- Encourages recycling of by products.
- Generates income round the year.
- Solves energy and fodder crisis.
- Generates employment.
- Provides opportunity for agro-based industries.
- Improves input use efficiency.

Key barriers

- Availability of financial resources.
- Access to information, extension, skills.
- Varying conditions on farmers fields.

Integration of enterprises tend to be more profitable in most of the cases. Employment potential of integrated farming system is higher than arable farming. These systems may help in alleviating poverty and improve the economic conditions and living standard of rural population in developing countries. More emphasis is still required to generate a generalized model suited to various farm size holdings in different agro climatic conditions. Research is needed to identify appropriate component combination of the integrated farming system for different situations.

Poverty impacts

- Sustained growth in agricultural sector is necessary.
- Growth in agriculture is more poverty reducing than growth in other economic sectors.
- Lack of capital assets such as human-, natural-, physical-, social- and financial capital.
- People's vulnerability, powerlessness and aversion to risks.

❐❐❐

Chapter-9

Land Use Systems and Agroforestry

Agroforestry involves multiple land use production system. But, land is used for agroforestry or growing of arable crops according to its capability. The ability of land is determined on the basis of (i) inherent potentialities, (ii) external land features, and (iii) environmental factors. In fact, soil properties like soil texture, soil depth, slope and extent of erosion are considered while classifying land capabilities. The land is classified into eight classes. From class I to IV are fit for arable crops whereas V to VIII classes are used for non-arable crops (for fuelwood, fodder or pastures). Usually, class VIII is used for recreation, wildlife, water supply or aesthetic purposes.

Agroforestry can successfully be taken up on arable lands. Selected tree species are synergistic and non-antagonistic.

The following are main landuse systems under agroforestry :

- Shifting cultivation
- Taungya system
- Woody perennial plantation
- Plantation forestry
- Multiple cropping.

Shifting cultivation is a primitive method of cultivation which is usually followed in hills by tribals in which forest land is cleared by cutting down all the trees shrubs, herbs and climbers near ground level; felled material is allowed to dry and then set on fire; sowing is followed in cleared area by dibbling small holes in the ground by mean of wooden stick pointed at the extreme end or metal rod or seeds are broadcast.

The cropping is done a few years on the same unit of land and then left for the regeneration of the bush (fallow period). Clearing and burning of vegetation leads to a disruption of the closed nutrients cycle on the forested land. However, burning causes raising in soil temperature temporarily, more solar radiation is received on the bare soil surface resulting in higher soil and air temperatures. This change in the temperature regime causes changes in the biological activity in the soil. The addition of ash to the soil through burning causes important changes in soil chemical properties and organic mailer content. In general, exchangeable bases and available phosphorus increase slightly after bruning; pH values also increase, but usually only temporarily.

After clearing and burning, a sharp increase of available nutrients may benefit the first crop but later on the soil becomes less and less productive and crop yields decline. Addition of manure and fertilizers may maintain soil fertility but in shifting cultivation system application of nutrients is seldom. When raising of crops become non-remunerative, the cultivators abandon the site and move into other sites. After a lapse of a few years again the same site is used for cultivation. By that period cut tree regenerates and soil also regains in its fertility and productivity. However, the next return has low productivity then the previous turn.

In India, shifting cultivation is practiced in the north eastern hill region (Assam, Meghalaya, Manipur, Nagaland and Tripura, Arunachal Pradesh and Mizoram), Andhra Pradesh, Bihar, Madhya Pradesh, Orissa and Karnataka. Shifting cultivation is called "**Jhum**" in north eastern hill region and "**Podu**" in AndhraPradesh and Orissa states. The size of the plot varies from 1.0 to 2.5 hectares per family having members 3 to 5.

Demerits of shifting cultivation

- Shifting cultivation is a faulty land use. It enhances soil degradation by way of soil erosion. Shifting cultivation is a social evil.
- It is a primitive method of cultivation in which crop yields are less and requires hardwork.

- Due to complete cutting and burning of vegeration there is an abrupt changes in soil microbial population.
- It causes environmental degradation.
- There is loss of nutrients from the soil through run off, leaching and percolation.
- The cation exchange capacity of soil is markedly reduced. The carbon, nitrogen ratio reduces due shifting cultivation.
- The excessive run-off in the shifting cultivated areas causes floods in the adjoining plains.
- Denudation of hill slopes takes place.
- Wild life in the shifting cultivated areas is badly affected. It disturbs the fauna.
- Forest flora become weak and grow sparsely.

Ways for improving degraded areas caused by "Shifting cultivation" during fallow period

The loss of soil fertility and reduction of productivity are bound to occur in shifting cultivation, but the following ways can restore and maintain level of productivity.

- Efficient land use management including provison for horticultural crops, forestry crops preferably legume species and livestock like cattle, sheep and goats.
- Preventing loss of soil fertility by making contour bunds, graded bunds, halt moon teraces, levelling, drainage system and water utilization technology.
- Soil management by growing cover crop, strip and mixed cropping, erosion checking crops, relay cropping, green manuring and use of organic amendments and fertilizers.
- Adoption of production based crop management by raising high yielding varieties, suitable crop planning, weed and water management, timely plant protection measures, application of fertilizers for restoring soil fertility and use of implements.
- Adoption of soil enriching hedgerow intercropping, or 'alley cropping' aims at eliminating the fallow period altogether by combining tree species with agricultural crops. The problems created by shifting cultivation can easily be overcome by adoption of suitable agroforestry models.

Taungya System

The 'Taungya' is a Burmese word, consisting of '**Taung**' means hil and '**ya**' means cultivation i.e. cultivation in the hills. Taungya is also a shifting cultivation in which cultivators are allowed ot plant forest tree species and to raise agricultural crops in between rows of forest tree species simultaneously, and cared for about 3 to 5 years. Thereafter, cultivators are required to shift to another patch of land. The aim of Taungya system is both wood and food production.

Taungya is locally called **kumri** (Tamil), **ponam** (Malayalam) **podu** (Telugu and Oriya), **Jhum** (Assamese) etc. Taungya was reported to have started first in Burma in the year 1850 for replanting vast areas under teak plantation, and in Java in 1856. In fact, it is a means of raising forests. Taungya was introduced to India by Brandis, in 1856. The first plantations were raised in 1863 in North Bengal. It is practiced in Kerala, West Bengal, Uttar Pradesh, Tamil Nadu, Andhra Pradesh, Orissa, Karnataka, and North eastern hill regions. This system provides employment opportunities to unemployed rural people. It is also suitable in the area where land shortage is very acute. In India, Taungya is followed in Bihar where adequate amenities are provided to villagers like schools, co-operative forest societies, health centres, facilities for drinking water. In Uttar Pradesh, widely spaced tree species are permitting agricultural crops to grow successfully.

Types of Taungya

Departmental Taungya

This is done by forest department. The labourers are employed on daily wage basis. The agricultural crops are grown mainly for suppressing weeds between the rows of tree species.

Leased Taungya

The land is leased by cultivators from forest Department, and crops are raised in between rows of the tree species. Sometimes, it is also done after the felling of sal trees as it is done in Uttar Pradesh. There is proper regeneration of forest as cultivation is done for growing crops which controls weeds also.

Village Taungya

In the forest area villages are settled under the superviision of Forest Department. Villagers are alloted about 0.2 to 0.4 ha or so piece of land per family for 3 to 4 years at one place. Villagers clear the land by collecting stumps and debris, and thereafter they cultivate the land, and raise crops.

Taungya cultivators are looked for weeding and hoeing, watering, tending and thinning of seedlings, protection from grazing, watch and ward, fencing etc.

Main crops of Taungya

Jowar, bajra, paddy, maize, lentil, wheat, mustard, rapeseed, gram, arhar, urad, soybean, castor, sunnhemp, barley, potato, sweet popato tapioca, til, groudnut, turmeric, ginger etc.

Crops grown in taungya system in India

State	Tree crop	Associated agricultural crops
Uttar Pradesh	*Shorea robusta, Tectona grandis, Acacia catechu, Dalbergia sissoo, Eucalyptus spp., Populus spp.*	Maize, paddy, sorghum, pigeon pea, soybean, wheat, barley, chickpea, rape seed and miscellaneous.
Andhra Pradesh	*Anacardium occidentale, (cashew), Tectona grandis, Bombax ceiba, Bamboo, Eucalyptus spp.*	Hill paddy, groundnut, Sweet patato.
Kerala	*Tecona grandis, Bombax ceiba, Eucalyptus spp.*	Paddy, tapioca, giner, turmeric
Assam	*Shorea robusta*	Paddy
Tamil Nadu	*Tectona grandis, Bamboo, Santalum album, Tamarindus indica, Acacia mearnsii, Ceiba pentandra, Cashew, Rubber*	Millets, pulses, groundnut, cotton, tapioca, potato
West Bengal	*Tectona grandis, Shorea robusta, Schima wallichii, Cryptomeria japonica, Quercus spp., Michelia doltsopa*	Paddy, maize, millets, turmeric, ginger, okra, pineapple, hemp
Andaman and Nicobar Islands	*Pterocarpus dalbergioides*	Sugarcane, maize

Merits of Taungya

- Weed and climbers growth is suppressed.
- Regeneration of forest species is cheap.
- Employment opportunities to unskilled unemployed rural people.
- There is full utilization of available land for the production of food crops.
- This is an important way to integrate rural devlopment programmes.
- Increases forest wealth of the country.

Demerits of Taungya

- Exposure of land leads to erosion and loss of soil fertility.
- There is an exploitation of human labour.
- Danger of epidemics is expected due to raising of agricultural crops.
- Forest trees are not adequately cared by cultivators once they are settled.
- Cultivators may claim for legal right on land.
- Insecure land tenure.

Woody Perennial Plantation

The woody perennial plantations are coconut, oil palm, arecanut, cocoa, coffee, tea, rubber, cashew etc. In these plantations, lot of land remains uncoverd by the canopy of plantation crops when they are preferred to grow as monoculture which leads to soil erosion and deterioration in soil fertility due to high rainfall which occurs for longer period of the year. It is appropriate if cover crops (mainly legume crops) are grown. The presence of more plant cover on the soil surface will also reduce erosion hazards and ensure better efficiency in the utilization of the native and applied nutrients.

Plantation Forestry

Raising of forests mainly for wood production, They are small in size. Plantation forestry is an economic land use activity. Specific plant species are planted for specific purpose in tropics. These are planted with annual crops like ginger, turmeric, tapioca, pineapple etc. The proper soil management will help in restoring fertility.

Multiple Cropping

Multiple cropping is a system of cropping in which more than one crop is grown on the same unit land in a year without any gap. The main aim of multiple cropping is to make effective use of modern farm technology and farm input with minimum cost of cultivation and without affecting soil fertility.

Type of Multiple cropping

- Sequential cropping where two or more crops are grown one after the other in short succession on the same piece of land; and
- Intercropping where two or more crops are grown at the same time on a piece of land. Intercropping intensifies farming in both space and time. Sequential cropping in time only.

Pre-requisites of multiple cropping

- Land holding
- Type of land in terms of its productive potential
- Irrigation facilities
- Economic position
- Locality
- Market

The following are a few examples of multiple cropping. The most successful cropping system of South India is coconut- cocoa-pineapple-black pepper. In north India, a few multiple cropping can be taken such as.

- Upland rio rice-wheat
- Maize - wheat - mung
- Upland rice-lentil
- Mung - maize - wheat - mung.
- Upland rice-gram

Advantages of multiple cropping

- In agroforestry, multiple cropping system is a remunerative proposition.
- Available resources (land, water, light, man power, bullock power, machinery and finance, can be utilised efficiently throughout the year.
- Multiple cropping generates employment to unskilled rural poor throughout the year.
- Total out turn per unit area and unit time is increased.
- A considerable addition of organic matter in the soil increases the soil fertility and improves the physical properties of the soil.
- Soil is protected form erosion losses.
- Raising of crop after crop controls the weed population.
- There is better utilization of soil fertility.
- There is better care of tree species along with agricultural crops.

Disadvantages of multiple cropping

- Multiple cropping system requires assured irrigation facilities.
- It requires only short duration, quick growing varieties.
- Its success depends on adoption of modern technology which is not available with the farmers.
- The poor financial condition of cultivators does not allow them to purchase all requirments for raising successful crops.

Guide lines for multiple cropping

- Know intial fertility status of soil through laboratory tests.
- Prepare a schedule of raising crops.
- Apply recommended doses of manure and fertilizers to each crop.
- Apply micronutrients like Zn, B. and Cu in the soil and/or spray on crops as and when it is needed.
- Include legume crops which fix elemental nitrogen.
- Select early maturing and high yielding varieties of each crop.

Intercropping in fruit crops (Agri - horticulture system)

It is well known that fruit crops give economic yields after 5-6 years of transplanting and many others take even 8-10 years or so. Only a few, are short duration like banana and papaya. Under such circumstances, it becomes essential to think on how to get additional income from the same orchard so that maintenance expenditure and cost of planting materials are obtained, if not getung additional profit at least. The solution to this problem lies only with raising of suitable intercrops on space available between rows of fruit trees i.e. agri-horticulture system. The present state of knowledge is based on relevant research work done in various corners in India.

Suitable intercrop combinations in fruit orchards

Name of main fruit crops	Intercrops
Mango	Onion, Chillies, Mung, Mash, Moth, Cowpea, Cotton, Peas, Turnip, Cauliflower, Carrot, Radish and Gram.
	Phalsa as filler in Mango.
	Knoll-Khol, Cabbage, Cauliflower, Radish, Turnip, Ginger and Pepper
	For non bearing orchards:- Soybean, Pea, Cowpea, Palak, and Chillies. For bearing orchards : Tomato, Cowpea, soybean and Coriander.
	Two long duration i.e. Banana and Tapioca and Two short duration i.e. Tomato followed by Clusterbean and Brinjal followed by Cowpea
	Vegetables like, Potato, Brinjal, Onion, Okra, Chillies, Tomato, Colocasia, Bitter gourd, Indian squash, Cauliflower and Long melon etc, Pulses like, Cowpea, Black gram Green gram, Pigeonpea, Pea and Gram etc.
Sapota	Tomato, Peas,Turnip, Radish, Beans, Mung, Mash, Gram, Cowpea. Besides, Peach, Plum, Phalsa and Papaya may be planted as "filler" crop
	Cowpea and French bean.
	Chillies, Tomato, and Cabbage etc. Besides intercrops of Turmeric, Ginger etc. are also grown in rotation.
Guava	Cowpea and French bean.
	Okra, Onion, Turmeric, Garlic, Cauliflower, Cabbage, Chillies, and some leafy vegetables.

Citrus	Berseem – Lucerne and Methi (an eight-year-old citrus orchard).
	Cotton
	Summer season : Onion, Chillies, Mung, Mash, Moth, Cowpea and Cotton (up to 3-4 years age) and Guar
	Winter season : Peas, Turnip, Cauliflower, Carrot, Radish and Senji and Gram.
	Vegetables and leguminous crops (Pulses) may be grown in proper rotation for few years.
	Guar and Senji.
	Tinda and Kaddu, Mung, Moth and Mash, Cotton, Guar, Peas and Gram and Senji.
	Leguminous crops.
	Peas, Lobia, Guar, Tomato, Okra and Potato.
	Pea, Cowpea, Gram and Beans.
	Okra, Cauliflower, Tomato, Brinjal and Cabbage.
	Potato
Coconut	Ginger
	Groundnut and Colocasia
	Tapioca and Elephant yam
	Elephant yam, yam and Tapioca
	Sugarcane
	Followed by Colocasia and Cassava and rice, Groundnut
	Cocoa
	Turmeric
	Banana
	Citrus species
	Clove
	Turmeric
	Elephant foot Yam, Turmeric, Ginger, Cassava and Banana
	Banana
Arecanut	Black pepper
	Banana, Pepper, Elephant foot Yam, Pineapple, Cacoa, Guinea grass
	Elephant yam + Black pepper, Arrowroot + Pepper, Betelvine, Pepper, Pineapple and Banana

	Turmeric, Sweet potato and Cowpea
	Banana
Vineyard	Field Crops :- Guar, Lentil, Green gram, Mash, Gram and Peas
	Vegetable Crops :- Tomato, Spinach, Beans, Chillies, Ginger, Turmeric & Sweet potato
	Flowers :- Marigold, Tuberose, Gladiolus and Jasmine
Banana	Brinjal, Colocasia, Yam, Chillies and Okra
	Ginger, Colocasia and Turmeric
	Onion, Tomato, Capsicum or Potato
	Turmeric, Ginger, Pineapple and some leguminous crops
	Turmeric
	Cowpea, Sweet potato and Maize
	Urad, Mung, Soybean, Groundnut or Cowpea
	Turmeric or Yam
	Lucerne, Sunhemp and Coriander
Papaya	Chillies, Onion, Tomato and Beans
	Chillies, Onion, Cabbage, Radish, Knol-Khol and Tomato
	Tomato, Chillies, Beans and Watermelons

Non-suitable intercrop combinations in fruit orchards

Name of main fruit crops	Intercrops
Mango	Sugarcane
	Wheat, Oat, Jowar, Bajra, Ginger, Colocasia and Turmeric
Guava	Soybean and Green gram
Citrus	Jower, Bajra, Maize, Millets etc.
	Sugarcane
	Maize, Papaya, Banana, Castor, Wheat and Rice
	Maize, Sugarcane, Cotton and Wheat and Creeping types of Vegetables
	Tobacco, Okra, Tomato, Brinjal and Chilli
	Berseem
	Wheat and Maize
	Paddy

	Chillies and Cotton
	Maize and Radish
Coconut	Tapioca, Rice and Ragi
	Sweet potato and Colocasia
Arecanut	Banana, Ginger, Turmeric and Pineapple
Banana	Sweet potato

❑❑❑

Chapter-10

Desirable Characteristics of Tree Species for Agroforestry

While selecting tree species for agroforestry systems, the following desirable characteristics should be taken into consideration. Though all desirable characters are not found in a single species, but their multiple uses are taken care of :

- Tree species should not interfere with agricultural crops with respect to soil moisture, nutrients, solar energy and agricultural operations, and should not attract birds.
- Tree species should have high survival percentage and easy establishment.
- Tree species should have fast growing habit and easy management.
- Tree species should have easily decomposable leaves.
- Tree species should have ability to fix atmospheric nitrogen.
- Tree species should not have any toxic effects on soil and on associated crops.
- Tree species should have adequate shade regulation, upright growth, respond to pruning, pollarding and coppicing.
- Tree species should have their multiple uses.
- Tree species should have high yield potential and higher profitability.
- Tree species should have wide acceptability by local people.

The following is the brief explanation on each characteristics for better understanding.

Tree Species should not Interfere with Agricultural Crop for

- Soil moisture
- Nutrients
- Solar energy (sunlight)
- Agricultural operations
- Should not attract birds

Tree species should not interfere with soil moisture

The water requirement of tree species should not be very high otherwise it will compete with main agricultural crops for water. Tree species should be deep rooted so that they can draw water from deep strata of the soil. Some *Eucalyptus spp.* reported to have very high water requirements. They should not be preferred for agroforestry. If, at all, it becomes necessary to include such species they should only be planted at wider spacing, and on farm bunds and boundaries of paddy crop.

Tree species should not compete for plant nutrients

Forest species should not utilize more plant nutrients rather they should help in building soil fertility, like leguminous tree species which fix atmospheric nitrogen in their roots. The root system and root growth characteristics should idealy result in exploration of soil layers that are different to those being traped by agricultural crops.

Tree species should not compete for solar energy (sunlight)

Tree species should not interfere with sunlight falling on the crops. However, sometimes it is advantageous to reduce intense sunlight, for this, trees play important role. Tree species should be light branching in their habit. Such trees, permit the penetration of light to the ground and promote better crop and pasture growth and yield. Trees having spreading branches and developing dense canopy interfere with sunlight. In such species regular pruning becomes essential, but species should be such, that can withstand pruning operations. There are well known tree species which cannot withstand regular pruning *viz.* mango, tamarind, jamun, jackfruit etc.

Trees should not interfere with agricultural operations

Tree species should not have prominent thorns otherwise carrying out agricultural operations will be difficult. For example *Prosopis juliflora* (*Vilayati babul*) plants are thorny which should be avoided in agroforestry or they should be trained and pruned regularly.

Trees should not attract birds

If tree species attract more number of birds, they will damage agricultural crops. In fact, trees bearing good fruit attract birds. They gather on them and eat the fruits, and also damage field crops. Banayan trees bear fruits profusely such tree species should not be selected for agroforestry systems particularly in Agri-silviculture system.

Tree Species should have High Survival Rate and Easy Establishment

Trees species having high survival percentage leave little or no gaps after transplanting. Hardy tree species are easy to establish. They have less mortality percentage because they can tolerate transplanting shocks easily. Moreover, many of them have the ability to regenerate lateral roots within a short period of time after transplanting.

Tree Species should have Fast Growing Habit and Easy Management

Suitable tree species for agroforestry system should be essentially a fast growing which can grow beyond the reach of livestock at the end of the first monsoon season. The following species are fast growing.

Acacia auriculiformis
Acacia decurrens
Adenanthera pavoni
Anthocephala indicus
Bignonia megapotamica
Bombax ceiba
Cassia fistula
Casuarina equisetifolia
Cordial sebestina
Delonix regia
Eucalyptus citriodora
Kigelia pinnata
Kydia calycina
Leucaena leucocephala
Millingtonia hortensis
Muntingia calubura

Tree Species should have Easily Decomposable Leaves

The suitable tree species for agroforestry will be that one in which fallen leaves decompose easily. The leaves of most of the legume tree species are small in size, decompose quickly and easily, and add a large quantity of organic matter and nutrients to the soil. As far as possible, tree species having broad leaves should not be preferred for agroforestry system because they contain more fibre matter and also require longer time for decomposition. Further, broad leaves when fall on the tender crop plants block their photosynthetic activities. A few broad leaved tree species are teak, banayan, mango, cashew, etc. are not preferred over other suitable tree species for planting in agroforestry systems.

Tree Species should have Ability to Fix Atmospheric Nitrogen

The selected tree species should have capacity to tolerate dry conditions for growing in arid areas and so also they should be able to grow under high rainfall where they can tolerate water logging as and when it occurs. They should also tolerate saline and alkaline soil conditions which are very common, due to this, wastelands are increasing year after year. The following are a few tree species which help in fixing elemental nitrogen in their roots.

Acasia albida
Acacia auriculiformis
Acacia catecha
Acacia aneura
Acacia dealbata
Acacia decurrens
Acacia farnesiana
Acacia ferruginea
Acacia implexa
Acacia leucophloca
Acacia mearnsii
Acacia melanoxylon
Acacia mollissima
Acacia nilotica
Acacia planifix
Cassia fistula
Cassia siamea
Casuarinas equisetifolia
Dalbergia latifolia
Dalbergia sissoo
Delonix regia
Glyricidia sepium
Hardickia binata
Leucaena leucocephala
Moringa oleifera
Oogeina oojeinensis
Parkinsonia aculeata
Peltophorum ferrugineum
Pithecellobium dulce
Pongamia pinnata

Acacia senegal
Acacia nilotica
Albizia chinensis
Albizia lebbeck
Albizia procera
Alnus nepalensis
Alnus nitida
Bauhinia purpurea
Bauhinia variegate
Butea monosperma
Cajanus cajan
Prosopis alba
Prosopis chilensis
Prosopis cineraria
Prosopis jiliflora
Robinia pseudacacia
Samanea saman
Saraca indica
Sesbania aegyptica
Sesbania bispinosa
Sesbania grandiflora
Tamarindus indica

Tree Species should not have Any Toxic Effects on Soil and on Associated Crops

Tree species should not release any chemical (s) through their roots or through their leaves, which have toxic effects on soil and agricultural crops. Otherwise productivity of land will be reduced due to inadequate availability of nutrients. *Eucalyptus* leaves have been reported to contain certain organic acids and the soil under tree canopy absorbs these acids from dried leaves and turn acidic. Further, suppression of microbial population is also believed under the canopy of *Eucalyptus* tree. Tamarind is another such species.

Tree Species should have Shade Regulation, Upright Growth, Respond to Pruning, Pollarding and Copping

As far as possible, selected tree species should have upright growth, respond to pruning, pollarding and coppicing. Tree species which can withstand heavy pruning operation, will be able to regulate shade efficiently because their photosynthetic efficiency does not decrease with heavy pruning. Small crowned trees are better for shade regulation.

Tree Species should have their Multiple Uses

The selected tree species should have multiple uses. As far as possible, the tree should yield more than one of the main produce like fuelwood, leaf fodder, edible fruit, edible flower and fibre. Such raw materials will also be useful for establishing small rural cottage industries. Product diversification can avoid the glut in the market and prevent unhealthy competition among the cultivators.

Tree Species should have High Yield Potential and Higher Profitability

High yield potential is the most important criterion of selection of tree species for agroforestry systems as the main aim is to obtain overall more output per unit area. Care should be taken before collection of seeds and seedlings that they are being procured from reliable source.

Tree Species should have Acceptability by Local People

As has already been mentioned the choice of species depends largely on local climatic and edaphic conditions. These species should meet demands of local people, say, the species should directly or indirectly be related to socio economic pattern of the people. Sometimes, villages have their own choice for specific plant species because of their religious base and some other sentiments. Therefore, there must be full participation of local people so that while selecting plant species for formulating agroforestry models their interests are saved.

Characteristics of agrilcultural crops

- Agricultural crops should be short duration and quick growing.
- They should be at least partial tolerant to shade.
- Most of them should belong to leguminaceae family.
- They should respond well to high density planting.
- They should bear some adverse conditions like water stress and/or excess of watering; these conditions are inevitable time to time during growing season.
- Crops should return adequate organic matter to soil through their fallen leaves, root system, stumps etc.
- Crops should appropriately be fitted in intensive or multiple cropping system.

Dryland crops suitable for arid and semi-arid regions of India under agroforestry.

Cereal crops

- *Pennisetum glaucum*
- *Sorghum bicolor*
- *Zea mays*
- *Eleusine coracana*

Annual grain legumes

- *Cajanus cajan*
- *Cicer arietinum*
- *Cyamopsis tetragonoloba*
- *Vigna aconitifolias*
- *Vigna mungo*
- *Vigna radiata*
- *Vigna unguiculata*
- *Pisum sativum*

Oilseed crops

Arachis hypogaea
Brassica compestris
Carthamus tinctorias
Glycine max
Helianthus annuus
Ricinus communis
Sesamum indicum
Guizotia abyssinica

Fodder crops

Cyamopsis tetragonolaba
Pennisetum glaucum
Sorghum bicolor
Vigna aconitifolia
Vigna unguiculata
Zea mays

Grass species suitable for agroforestry

Aristida species	dabholu, Lapolu
Bracharia mutica	Para grass
Cenchrus biflorus	Anjan grass
Cenchrus cilliaris	Anjan grass, Dhaman
Cenchrus setigerus	Anjan grass, Dhaman
Chloris virgata	Rhodes grass, Gharnigha

Cymbopogon martini	Roich ghas, Rosha ghas
Cynodon dactylon	Dub
Cynodon plectostachyum	East African Star grass
Dichanthium annulatum	Marvel grass
Deplechnica fusca	Karnal grass
Eragostis curvula	Weeping love-grass
Eulatiosis binata	Baggar
Eremopogon foveolatus	Ghandel
Eleusine flagellifera	Chhimber
Hetetopogon contourtus	Spear grass, Pilighus, Shurval
Lasiurus sindicus	Lasisurus
Panicum antidotale	Blue panic, Dhus ghas
Sehima nervosam	Sain, Sedva
Sporobulus helvolvus	Usar ghass
Sporobulus tremulvus	Usarghass
Sporobulus marginatus	Usarghass
Saccharum munja	Munj or Sarkanda

❑❑❑

Chapter-11

Sustainable Agriculture and Intercropping

Concept of Sustainable Agriculture

a) **Sustainable Agriculture** - "Sustainable agriculture is the successful management of resources for agriculture to satisfy changing human needs which maintaining or enhancing the quality of the environment and conserving natural resources" by CGIAR, 1988.

b) **Goals of Sustainable Agriculture** - Sustainalbe agricultural systems must maintain or enhance -

  i) Biological and economical productivity of crops,

  ii) Enhance the efficiency of use of input,

  iii) Lesser adverse environmental impacts both on and off the farm,

  iv) Minimize adverse environmental impacts on adjacent and down stream environments,

  v) Minimize the magnitude and rate of soil degradation and to enhance soil quality minimum adverse impact on soils and environment and

  vi) Enhance compatibility with social and political conditions.

c) **According to Gips 1986, Sustainable Agriculture is -**

  i) **Ecologically sound** - Which means that the quality of natural resources is maintained and the vitality of the entire agro-ecosystem from humans, crop and animals to soil organism is enhanced. This is best ensured when the soil is managed and the health of crops, animals and people is maintained through biological processes (self-regulations). Local resources are in a way that minimises losses of nutrients, biomass and energy, and avoids pollutions. Emphasis is on the use of renewable resources.

ii) **Economically viable** - which means that farmer can produce enough for self-sufficiency and/or income and gain sufficient returns to warrant the labour and cost involved. Economic viability is measured not only in terms of direct form produce (yield) but also in terms of functions such as conserving resources and minimizes risks.

iii) **Socially just** - Which means that resources and power are distributed in such a way that the basic needs of all members of society are met and their rights to land use, adequate capital, technical assistance and market opportunities are assured. All people have the opportunity to participate in decision-making, in the field and in the society. Social unrest can threaten the entire social system, including its agriculture.

iv) **Humanae** - Which means that all forms of life (plant, animal, human) are respected. The fundamental dignity of all human values as trust, honesty, self respect, cooperation and compassion. The cultural and spiritual integrity of the society is preserved and nurtured.

v) **Adaptable** - Which means that rural communities are capable of adjusting to the constantly changing conditions for farming: population growth, policies, market demand etc. This involves not only the development of new, appropriate technologies but also innovations in social and cultural terms.

Role of Sustainable Agriculture

- Towards Low External Input and Sustainable Agriculture (LEISA)
- Towards Sustainble Agro-ecosystems
- Indices of Sustainability
- Towards Low Extermal Input and Sustainalbe Agriculture (LEISA) - LEISA is an option which is feasible for a large number

of farmers and which can complement other forms of agricultural production. As most farmers are not in a position to use artificial inputs or can use them only in small quantities, it is necessary to concentrate on technologies that make efficient use of local resources. Also, those farmers who now practice HEIA could reduce contamination and costs and increase the efficiency of the external inputs by applying some LEISA techniques. It is important that the agro-ecological knowledge of both scientists and farmers be applied, so that the natural resources are conserved and enhanced, productivity and security are increased and negative environmental effects are avoided.

LEISA Refers to those forms of Agriculture that

Seek to optimize the use of locally available resources by combining the different components of the farm system, i.e. plants, animals, soil, water, climate and people, so that they complement each other and have the greatest possible synergistic effects.

Intercropping (Aiyer, 1949)

Intercropping is defined as growing of two or more than two crops simultaneously on the same period of time in the same field in rows of definite geometrical pattern. In rainfed areas intercropping is an insurance against bad weather. Selection of intercrop is generally from oilseeds and legume crops time and space dimension crop intensification. Intercropping in pearlmillet generally increases the land equivalent ratio (LER - Willey and Osiru, 1979).

Mixed cropping/Intra cropping (La-Flitze, 1928)

Mixed/intra cropping is the cultivation of two or more than two crops simultaneously on the same piece of land in without any definite row pattern or fixed ratio.

On the basis of per cent of plant population used for each crop in intercropping system; intercropping is of two types viz., (i) additive series and (ii) replacement series.

(a) Additive Series Intercropping

In such type of intercropping, one crop is main crop or base crop and another crop is intercrop. Intercrop is introduced into the base crop by adjusting or changing crop geometry i.e. addition of intercrop to the base main/crop. Here plant population of base crop is same to what recommended population in pure stand whereas that of intercrop is less. Cropping husbandry is according to the base crop. this type of intercropping is prevalent in India and its main objective is to get additional income and to cover risk eg-sowing of potato in the field of sugarcane in between the rows of crop at a 90 cm.

Sowing of seeds is generally done by broadcasting method commoly practised in Dryland area of India. It is subsistence farming. Main objective is to lessen the risk of total crop failure, and to satisfy the farmers in food and fodder. Intercropping is the new concept of mixed cropping. Scientific study of mixed cropping was firstly done by La-Flitze (1928).

(b) Replacement Series Intercropping

Both the crops are component crops. Neither is the base crop nor is the intercrop. It means the plant population of both component crops is less than their recommended population in pure stand. Such type is widely practised in western countries.

By growing of the above three intercrops individually with sugarcane, the production of none is hardly affected.

Difference between mixed (intra) and intercropping

Mixed/Intracropping (La-Flitze, 1928)	Intercropping (Aiyer, 1949)
• At least one single crop can survive at adverse weather condition.	Proper utilization of inter space in the base crop.
• None of crop consider primary and secondary. All are equal for their management practices.	Main crop is cared much better than inter crop (in additive series).
• Competition between the all crop is equal.	Intercrop have not completion with main crop eq-potato + sugarcane.
• The harvesting/maturity of all the crops are same.	Intercrop mature earlier than the base crop.
• Sowing time of all the crops are generally same.	There is time gap in between the base and inter crop for sowing is same or different.
• Sowing can be done by broadcasting method.	Only line sowing is possible.
• The seeds of different crops are mixed before sowing.	Mixing of seed is nor possible for line sowing.
• There is a problem in harvesting because clasping the stem by tender plant (cowpea + pearlmillet).	Harvesting is easy. Separate line and no clasping.

Pre-requisites of Successful Intercropping

The main objective of intercropping is higher productivity per unit area in addition to stability in production. Intercropping utilizes resources efficiently. There are certain important pre-requisites of intercropping:

1. The time of peak nutrient demands of component crops should not overlap eg-maize + Urd/Mung.
2. The peak nutrient demand is 30-35 day after sowing in case of Urd/Mung whereas it is 50-55 days after sowing (DAS) of maize.
3. Competition for light should be minimum among the component crops.
4. Complimentarily should exist between the component crops.
5. The differences in maturity of component crops should be at least 30 days.
6. Competition for CO_2 and H_2O should also be minimum among component crops.

Advantages of intercropping

- Better use of growth resources, including light, nutrients and H_2O.
- Suppression of weeds.
- Reduced plant and disease incidence.
- Yield stability.
- Ecological stability i.e. - improvement of soil health and agro-ecosystem.
- Other eg-spreading of labour, physical support of one crop to another and home gardening leading to a more pluriform food supply is a good example of realization of intercropping advantages.

Disadvantages of intercropping

- Labour intensive.
- Control of pests and diseases or chemical weed control may be difficult.
- Mechanization is difficult to realise.
- Disadvantages caused by adverse competitive effects or by allelopathy.

Choice of Crops for Multiple Intensive Cropping

1. The crops should be arranged in such a way that allelopathic effect, temporary immobilization of nutrients and depletion of nutrients from the same layer of soil do not occur.
2. Legumes have a definite place in any cropping system because of their limited demand on resources like water, nutrients and light, their adaptability to varying environmental conditions and capacity to fix atmospheric N in root nodules.
3. Vegetables deserve their due place in intensive cropping because
   a) High cash and nutritional values.
   b) Most of vegetables can be raised as seedlings and planted/ transplanted.
   c) Thus reducing main field duration and competition with the associated crop.
   d) They can be grown in interspaces in orchards and.
   e) Plantations crops.
   f) They can be inter or relay cropped in cereal base crops.
4. Short duration photoperiodically insensitive genotypes are suited for intensive cropping system.

Plant Interactions

(A) Interaction between component crops

The interaction is mainly due to response of one species to the environment as modified by presence of another species. Interaction may be competitive or non-competitive or complementary. If the crops are grown in association and the growth of either of the concerned species is not affected, such type of interaction is called as non-competitive interaction or interference. If one species is able to help the other it is known as complementary interaction if supply of N fixed by the legumes to the associated non-legumes. (This is otherwise known as annidation). One species may have greater ability to use the limiting factor and will gain at the expense of the other. This is called as competitive interaction or interference. Interaction may also occur in some other manner by way of producing toxic substances and affecting the establishment and growth of the associated species. This is called as allelopathy/teletoxy.

(B) Interactions in intercropping

Factors such as light, nutrient, O_2 and CO_2 are required for plant growth. The nature of such competition may vary depending on the density and proportion of each component and planting pattern.

i) **Light**- Photosynthetically active radiation (light) is directly related to photosynthetic rates of component crops. Intercropping can increase light interception by much as 30-40%. When one component is taller than the other in an intercropping system, the taller components intercept most of the solar radiation. The pattern of light interception also varies according to the age of the crop. Light transmission by coconut is only about 20% at the age of 8-10 years and this remains almost constant till about 25 years of age. Subsequently the % light transmission increases progressively as the canopy coverage of ground decreases. When the palms are about 40 years of age, light transmission increases to about 50%. In general, the component crops under intercropping situations are grown in such a way that competition for light is minimized.Proper choice of crops and varieties, adjustment of planting density and pattern are the techniques to reduce competition and increase the light use efficiency.

ii) **Moisture and Nutrients** - The aggressive component generally absorbs greater quantity of nutrients and soil moisture. In legume and non-legume combination, the later takes up large amount of P, K and S. As a result, the legume may show deficiency of these nutrients. Such effects, however, may be mitigated by appropriate fertilizer application. Among intercrops, sorghum and pearlmillet are more competitive in extracting nutrients. Generally the intercropped stand remove greater amount of nutrients than sole crop stands.

Allelopathy (Hans)

Green plants produce numerous secondary metabolites, may of which are capable of initiating chemical warfare among the neighbouring plants growing in a community. These chemicals have been designated as allelo-chemicals and the process as allelopathy. Both, crop and weedy plants may possess such allelopathic compounds. The allelopathic compounds studied so far are found to be largely derivatives of benzoic acids, cinnamic acids, phenolic acids, coumarins, hydroquinones, benzoquinones, thiopenes, juglone, and cinedes. Some authors prefer to call the harmful plant interactions of such chemicals as teletoxy, but the broader for allelopathy is still widely accepted.

Sunflower, sorghum, walnut, cucumber, peach and *Eucalyptus* are the examples of such crops which are known for allelopathic effect. It is following types–

1. **True Allelopathy** – The direct or indirect harmful effect on the other crops.
2. **Functional Allelopathy** – When precursor is released by micro-organism
   i) Auto-inhibition – Damage the producer species itself.
   ii) Allo-inhibition - Damage the species of crops.

Legume Effect

The beneficial effects of the legumes in any crop rotation and intensive cropping system is termed as legume effect. Inclusion of legumes in the cropping system is beneficial in many ways –

1. Legumes fix atmospheric nitrogen in root nodules and thus improve the nitrogen status of the soil.
2. It saves upto 25% of recommended level of nitrogen application to the associated cereals when grown as intercrop.
3. The crop residues and root nodules of legume release nitrogen during decomposition for the use of the succeeding crop.
4. Legumes absorb soil phosphorus more efficiently and part of this mobilised phosphorus in organic form is available to the succeeding crop.
5. Legume have greater root Cation Exchange Capacity (CEC) than cereals. Thus promotes Ca^{++} and Mg^{++} availability to plants.
6. Many of legume can tolerate some amount of shading and drought.

Sorghum Effect

The nutrient status of soil is exhausted by any fast growing cereals. The crop residue of such cereal having wide C : N ratio takes a long time to decompose. During the process of decomposition, soil nitrogen is temporarily immobilized affecting the succeeding crop. Such effect is more pronounced in sorghum in low fertile soil causing to temporary deficiency of nitrogen in the soil for the succeeding crop. To reduce sorghum effect, 25% more nitrogen is applied as the time of first fertilizer dose of the succeeding crop. It hastens the process of decomposition and overcomes the immobilised nitrogen. *Ragi* is the exception because its residues decompose rapidly resulting in mineralization of soil nitrogen.

Cotton Effect

Cotton feeds in the deeper layers of the soil, and removes comparatively smaller quantities of nutrients. The succeeding crop having the shallow root system is to tap on the unfed pool of nutrients in the surface layers of the soil. This effect may be termed as cotton effect.

Types of Intercopping

Intercropping includes

- Alley cropping
- Strip cropping
- Contour cropping
- Paired row cropping
- Skip cropping
- Parallel cropping
- Companion cropping
- Multi storeyed cropping and
- Synergetic cropping

Among the above the major intercropping systems are –

1. Parallel cropping

Cultivation of such crops which have different natural habit and zero competition eg-blackgram/ greengram + maize. The peak nutrient demand period for blackgram or greengram is around 30-35 days after sowing while it is 50 days after sowing of maize.

2. Companion cropping

Such intercropping where the production of both intercrops is equal to that of its solid (sole) planting eg-Mustard/Potato/Onion + Sugarcane.

3. Multi-storeyed / Multi-tiered / Multi level cropping

Cultivation of two or more than two crops of different heights simultaneously on a certain piece of land in any certain period of time eg. sugarcane + mustard + onion/potato.

4. Synergetic cropping

The yield of both crops are higher than of their pure crop on unit area basis eg- sugarcane + potato.

Alley Cropping

Alley cropping is an Agroforestry system in which fast growing N-fixing shrubs are planted as hedgerows. Food crops/arable crops are grown in alleys formed by hedgerow of trees and shrubs. It is also known as hedgerow cropping or hedgerow intercropping system. Alley cropping is an important step in meeting the fodder fuel and food requirement of the farmer. This system is most suitable for marginal and sub-marginal lands. Scientists at International Institute for Tropical Agriculture (IITA), Nigeria, have developed this system of Agroforestry in the 1970s.

In this system, trees or shrubs, preferably legumes are planted in rows along the contour. The hedgerows are periodically pruned to prevent shading and to reduce competition with the associated food crops. In the summer after the harvest of food crops, the hedgerows are allowed to grow freely to cover the land. Agronomic management of both the hedgerows and alley crops is very important to optimize the production.

Advantages

- Alley cropping is more remunerative under rainfed conditions.
- The trees are often pruned and lopped. The foliage obtained from these trees is either used as green manure or mulch in the field or as fodder for animal during lean period.
- In the semi-arid regions, where organic matter and nutrients are deficient, alley cropping has high potential for improving the soil fertility and increasing the biomass production. It has been reported that growing *Leucaena* enabled improving in the soil fertility.

- On an average, there was increase in organic C from 0.37 to 0.43%, available N from 163 to 190, P from 8.2 to 14.9 and K from 460 to 500 kg/ha with decrease in soil pH.
- In the hilly areas, alley cropping has specific role of protecting the soil from erosion. The tree hedge and alternate strips of erosion permitting and resisting crops reduce water and soil loss and increase the productivity of land.

Classification of Alley Cropping

According the purpose for which the alleys are raised, alley cropping may be grouped into :

i) Forage alley cropping,
ii) Green-manure-cum-mulch alley cropping,
iii) Forage-cum-mulch alley cropping and
iv) Forage-cum-green manure alley cropping.

Among them, forage alley cropping is recommended as an alternate land use system for semi-arid regions with main benefit of green fodder production during the dry season.

Management of alley cropping

Several the species are used as hedge species. They should have the ability of easy establishment, fast growing nature and high foliage productivity. They should be tolerant to pruning and ability to coppice vigorously. Trees such as *Leucaena leucacephala, Sesbania grandiflora, Glyricidia sepium* and *Cassia siamea* are used in this system. Among these species, *Leucaena* is preferred due to its easy management on heavy soils with less than 800 mm of rainfall. *Leucaena* performs better than any other perennial herbaceous legume.

Arable crops are grown in the interspaces between the rows of trees or woody shrubs. Crop such as maize, sorghum, pulses and oilseeds (particularly safflower and peanut) are grown as alley crops. Based on the classical work at IITA, Wilson and Kang (1981) reported that the alley cropping system based on *Leucaena* + maize was promising because, *Leucaena* tops maintained maize grain yield at reasonable levels with no additional N input on a sandy soils. Under semi-arid tropics with vertisols, maximum production was obtained from the alley cropping system of *Leucaena* and *Sorghum*. *Leucaena* can be grown with sorghum + pigeonpea (1:1) for higher biomass and net returns. In entisols with 3% slope under rainfed condition, *Brassia campestris* var. yellow sorson grown during winter season within the alleys of *Eucahyptus hybrids and Dalbergia sissoo* gave fairly good yield. Under Jhansi (UP) conditions, *Leucaena leucocephala and Sesbania sesban* as hedge crops and pearlmillet as alley crop provided higher LER.

Alley Width

Proper maintenance of alley width is important to get fairly high yield of alley crops. Alley width may vary from 4.5 to 6.0m for vegetative cover and higher fodder yield. Wider alleys of 10-15m may serve as vegetative barriers in conjunction with terrace level structures. Studies at Jhansi have emphasized the need to maintain 6-8m alley width so as to minimize yield reduction of alley crops. In *Leucaena* + Maize, wheat and greengram system, fodder and wood yields were higher in the narrow alleys while grain production was higher in the wide alleys. Growth and yield of the crops are suppressed when grown close to tree rows. As the distance between the field crops and hedge plants increased growth and yield performance of the crops have also increased. Poor performance of intercrops nearer to the hedgerows is mainly attributed to competition between field crops and tree species light and H_2O. However, hedge plants did not have any adverse effect on the yield of crops during the initial period of establishment.

Hedge management

The perennial vegetation is managed in such a way as to keep the competition between the hedge species and field crops at a minimum. To reduce shading effect of the hedgerows on companion crops, frequent pruning should be adopted. Height of the hedge and intensity of pruning affect the biomass of hedge species. The cutting height varies from 0.5-0.75m and approximately 4 cuttings can be adopted in a year at every three months internal. It has also been reported that fairly good yield of arable crop could be obtained with the alleys of *Eucalyptus hybrid and Dalbergia sissoo* when pruned twice (May and October), a year before sowing of crops. Several creal fodder crops like. *Sorghum*, pearlmillet, maize and teosinite were grown along with *Leucaena* both under managed canopy at a height of 50-100cm and also as a full grown tree. Under managed canopy, the productivity of the associated crops were generally reduced by 15-20%, whereas under non-managed canopy the yield reduction of associated fodder crops was 40-60% during rainy season and 30% during summer season.

Tree management

Tree canopy or tree cover has enormous effect on the under storey grown crops. The tree having dense canopy permits very little solar radiation to fall on the underneath crop to meet its light requirement. There are severed management options available viz., coppicing, pollarding, lopping and pruning.

(A) Coppicing

The word 'coppice' means the ability to regenerate by shoot or root suckers of an established forest trees. This practice involves regular harvesting of trees which produce numerous shoots from the stump after cutting. Coppice sprouts usually grow vigorously because they are served by roots big enough to feed the former tree.

(B) Pollarding

In this practice, the branches are removed usually at a height of about 1-3m at periodical intervals. This practice is mostly adopted for live fence and alley cropping.

(C) Lopping

During the lean period, major branches are removed to obtain fresh fodder. The fodder material obtained due to this practice depends on the species, its age, moisture level, soil fertility, season, severity of lopping and weather conditions. Lopping is a powerful management tool for controlling crown spread in the Agroforestry systems.

(D) Pruning

Small branches and twigs are removed one or two times during a year. The biomass/foliage obtained due to this operation is used as fodder. Such a practice is mostly adopted for maintaining trees/shrubs especially is live fence and alley cropping systems.

Ley farming

Ley farming or ley rotation is one of the alternate land use systems aimed to increase the food and fodder production and to enhance soil fertility concurrently. In this system, grasses and/or legumes are grown in rotation with arable crops. The objectives of this system are tonnage production of fodder and enriching soil fertility. Ley farming is more productive as well as sustainable against climate and economic risks than cropping done. Productivity of drylands can be improved by incorporating leguminous pastures/crop rotation. Soil fertility is increased when a ley is sown for 3 years or more. Considerable amount of organic matter is added through roots and above ground biomass and legumes. There is reduction in weeds, pests and diseases since ley provides a break in the crop sequence. There is improvement of nutritious leguminous fodder crops in drylands. Yield of grain and stover of sorghum increased when it is grown after four years of continuous cultivation of *Stylosanthes hamata.*

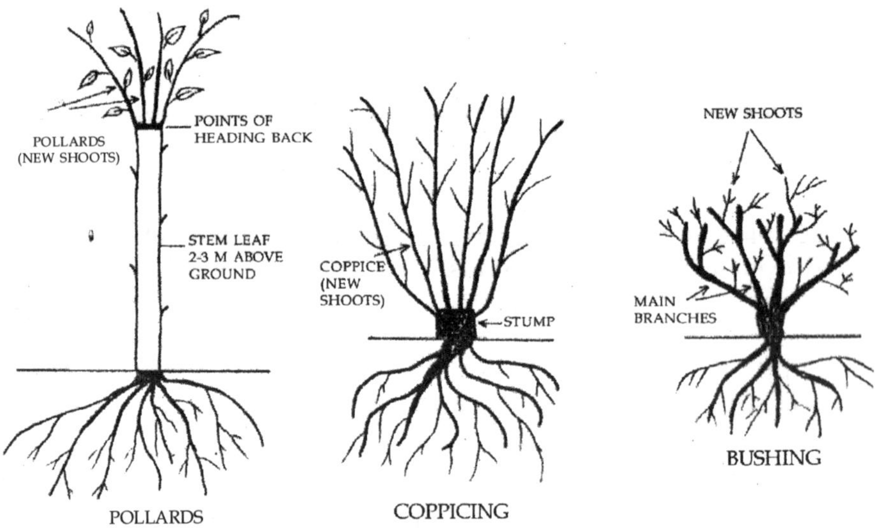

Chapter-12

Agronomic Indices of Various Farming System

1. Multiple Cropping Index (MCI)

It measures the sum of areas under various crops raised in a single year divided by net area available for that cropping pattern multiplied by 100. It is calculated for each cropping pettern separately and is very similar to cropping intensity :

$$\text{MCI} = \frac{\text{Total No. of crops} \times \text{Their respective areas}}{\text{Total area available for the cropping pattern}} \times 100$$

2. Diversity Index (DI)

It indicates the multiplicity of crops or farm products which are planted in a single year by computing the reciprocal of sum of squares of the share of gross revenue received from each individual farm enterprise in a single year. It may be calculated by using following equation :

$$\text{DI} = \frac{1}{\text{Sum of squares of gross revenue received from crop or enterprise in a single year}}$$

Diversity Index (DI)

DI was coine by Strout (1975) and Wang and Yu. It measures the multiplicity of crops or farm products which are planted in a year by computing the reciprocal of sum of squares of the share of gross revenue received from each individual farm enterprise in a single year.

Question 1. Compute the DI from the following farms and suggest which farm is most diversified and which is more special.

Name of crop	Income (Rs.) from Different Farms		
	A	B	C
Sugarcane	30,000/-	-	10,000/-
Cotton	10,000/-	20,000/-	20,000/-
wheat	40,000/-	20,000/-	10,000/-
Jowar	20,000	10,000/-	40,000/-
Potato	-	50,000/-	-
Total	1,00,000/-	1,00,000/-	80,000/-

Solution : Shares of individual crops in different form

Name of crop	Farm 'A'		Farm 'B'		Farm 'C'	
	Share	Squares of its share	Share	Squares of its share	Share	Squares of its share
Sugarcane	0.3	0.09	-	-	0.125	0.0156
Cotton	0.1	0.01	0.2	0.04	0.250	0.0625
Wheat	0.4	0.16	0.2	0.04	0.125	0.0156
Jowar	0.2	0.04	0.1	0.01	0.500	0.2500
Potato	-	-	0.5	0.25	-	-
Total	1.0	0.30	1.0	0.34	1.0	0.3437

Diversity index

$$\text{Farm 'A'} = \frac{1.0}{0.30} = 3.33, \text{ Farm 'B'} = \frac{1.0}{0.34} = 2.94, \text{ Farm 'C'} = \frac{1.00}{0.3437} = 2.91$$

Result : Therefore, Farm 'A' is more diversified and Farm 'C' is more specialed. Lower the diversity index, higher is the specialization.

3. Harvest Diversity Index (HDI)

It is calculated by using the DI equation except that the value of each farm enterprise is replaced by the value of each harvest.

4. Simultaneous Cropping Index (SCI)

It is calculatede by multiplying the HDI with 10,000 and dividing by MCI as stated below :

$$SCI = \frac{HDI \times 10000}{MCI}$$

5. Cultivated Land Utilization Index (CLUI)

It may be calculated by adding the products of land area planted to each crop, multiplied by the actual duration of that crop to reach physiological maturity and dividing by the total cultivated land area times 365 days :

$$\text{CLUI} = \frac{\text{Sum of the products from each crop} \times \text{Actual duration (days) of respective crop}}{\text{Total cultivated land area} \times 365}$$

It is applicable only in sequential cropping and the utility or sanctity of CLUI becomes very limited if the harvesting is delayed or in case of relay cropping and overlapping cropping systems.

6. Cropping Intensity Index (CII)

It determines the actual land use in area and time relationships for each crop or group as against the total available and area and time. It may be calculated by using following equation :

$$\text{CII} = \frac{\text{Total No. of crops grown (in unit time)} \times \text{Area under the crop} \times \text{Duration of that crop}}{\text{Farmers total cultivated land available for cropping} \times 365 \times \text{Total No. of fields temporarily available for cropping during study period} \times \text{Area of a specific period} \times \text{Time period for which that plot is available to farmer for cropping}}$$

7. Specific Crop Intensity Index

It may be calculated by using CII equation and determines the amount of area-time denoted to each crop or group of crops as against the total time available to the farmer.

8. Relative Cropping Intensity Index (RCII)

It is again a modification of CII which determines the amount of area-time allotted to one crop or group of crops as compared to the area-time actually needed in production of all the crops :

$$\text{RCII} = \frac{\text{SCII denominator}}{\text{CII numerator}}$$

If RCII is more than 50% for a specific crop then that farmer is said to be a specialized grower of that particular crop.

Statewise irrigated area (in%) and average cropping intensity in the 0.5 to 1.0 ha. size of holding

State	Net irrigated area (%)	Cropping intensity (%)
Andhra Pradesh	47.6	116.1
Assam	7.2	119.5
Bihar	25.9	133.6
Karnataka	20.9	106.8
Kerala	14.0	116.4
M.P.	14.5	123.4
Maharashtra	13.2	124.2
Orissa	19.6	141.5
Rajasthan	23.0	126.5
T.N.	43.6	122.6
U.P.	42.5	137.9
W. Bengal	17.5	127.6
All India	30.8	128.2

Source : Indian Agric. 15th Edition, Directorate of Eco. & Statistics, Min. of Agric. & Irrigation, Sept. 1980.

Evaluation of mixed/intra or intercropping system

Studies on the mutual relationship of plants in a mixed population and their relative profitability

Studies on individual plant types for their inter-relationship and respective profitability alongwith total productivity of the land are most essnetial so that intercropping or mixed cropping can be advocated with scientific reasoning and more convincing manners. Following studies would indicate whether such croppings are beneficial and should be adopted or not :

Aggressivity (A) :Studies on aggressivity show the relationship between dominant and dominated species grown together, especially when any of the two offers competition. It may be calculated by using the following equation :

$$\text{Aggressivity value} = + \frac{\text{Mixture yielf of 'a'}}{\text{Expected yield of 'a'}} \frac{\text{Mixture yield of 'b'}}{\text{Expected yield of 'b'}}$$

where,

A = aggressivity,

Aab = aggressivity value,

a = stand of one crop

b = stand of another crop.

An aggressivity value of zero indicates that both the species are equally competitive. Any other value will show difference in competitive values. A positive sign would indicate the dominant species and a negative sign the dominated one. Both positive and negative values would indicate a greater difference in actual and expected yield of both the species grown together.

Competition coefficient

This is studied with a view to find out the relative crowding of individual crop in a mixed population from which maximum yield can be obtained without any adverse effect on any of the species. It may be defined as the ratio of the relative crowding coefficient of any given spp. in the mixture. It may be calculated by using following equation :

$$\text{Competition of Coefficient (C)} = \frac{\text{RCC of given species}}{\text{Total RCC of all crops in mixture}}$$

Competition Index

It is studied with an object to find out the yield of various crops when grown together as well as separately. If the yield of any crop, grown together, is less than its respective yield in pure stand then it is said to be a harmful association but an increased yield a positive benefit and the crop should be grown together. Thus it indicates the yield per plant of different crops in mixture and in their respective pure stand on a unit area basis. The competition index may be calculated by the following equation (supposing wheat and gram are grown together)

$$\text{Competition Index (C.I.)} = (Ya - Ya^{+}) \times (Yb - Yb^{+}) / Ya\ Yb$$

where C.I. is Competition Index

Ya is yield of wheat/plant in pure crop of wheat

Ya^{+} is yield of wheat/plant in mixture

Yb is yield of gram/plant in pure stand

Yb^{+} is yield of gram/plant in mixture

The C.I. value of less than one indicates a positive benefit and the crops can be grown in association, whereas the values of more than one shows a harmful association.

Land Equivalent Ratio [(LER) Willey and Osiru, 1979]

It denotes the relation between land area under sole crop required to produce the same yield as obtained under a mixed or an intercropping system at the same management level. It is calculated as sum total of the ratios of yield of each component crop in an intercropping or a mixed cropping system to its corresponding yield when grown as a sole crop thus:

$$\text{LER} = \frac{\text{Yab}}{\text{Yaa}} + \frac{\text{Yba}}{\text{Ybb}}$$

Where, Yab is yield of species 'a' in association with species 'b' and Yba is the yield of species 'b' in association with species 'a' Yaa and Ybb represent the pure stand yield of species 'a' and 'b' respectively.

The crops with a lesser LER is dominated and with greater value is dominant one. Thus it indicates whether association is beneficial or not and which crop is more harmful so that the association of such crops may be avoided.

1. Suppose LER of peanut :
   Since, yield of groundnut intercrop = 1000kg
   Since, yield same sole crop = 1200kg
   Therefore, LER peanut = 1000/1200
2. Suppose LER of redgram:
   Since, yield of redgram intercrop = 600 kg
   Since, yield of the same sole crop = 1000kg
   Therefore, LER redgram = 600/1000

$$\text{LER of system} = \frac{1000}{1200} + \frac{600}{1000} + \frac{43}{30} = 1.43$$

Result: LER of 1.43 indicates that 43% yield advantage is obtained when grown as intercrops compared to growing as sole crops.

Yield of groundnut and pigeonpea as sole crops were 1000 and 500 kg/ha. Corresponding yield in intercropping was 700 and 400 kg/ha. Price of pigeonpea is Rs. 12/kg and that of groundnut Rs. 10/kg. Calculate LER and total monetary returns due to intercropping.

Solution:

$$\text{LER of system} = \frac{700}{1000} + \frac{400}{500} = 1.5$$

Monetary advantage

Financial value of sole groundnut	= 1000 × 10
	= Rs. 10,000
Financial value of sole pigeonpea	= 500 × 12
	= Rs. 6,00
Financial value of intercropped groundnut	= 700 × 10
	= Rs. 7,000
Financial value of intercropped pigeon pea	= 400 × 12
	= Rs. 4,800
Monetary returns due to intercropping	= Rs. 7,000+Rs. 4,800
	= 11,8000

(3) Yield and LER of pigeonpea intercropped with sorghum

Crops	Yield (kg/ha)	LERs	Total LER	Intercrop benefit
Sole sorghum	3952			
Sole Pigeonpea	1699			
Intercrop sorghum	3804/3952	0.96		
Intercrop pigeonpea	850/1699	0.50	1.46	0.46

(4) Yield and LERs of Pigeonpea intercropped with maize at 1:1 and 2:1

Proportion Crops	Yield (Kg/ha)	LERs	Total LER	Yield proportion of maize
Sole maize	3398			
Sole Pigeonpea	1035			
1:1 Intercrop Maize	2234/3398	0.66		
1:1 Intercrop Pigeonpea	896/1035	0.857	1.53	0.66/ 1.53=0.43
2:1 Intercrop Maize	3130/3398			0.92
2:1 Intercrop Pigeonpea	571/1035	0.55	1.47	0.92/ 1.47=0.63

Note: Now it misleading to argue that intercropping 1:1 is better than intercropping 2:1 proportion on basis of higher LER, Intercropping 2:1 may be preferred if the farmer is desirous to obtain a higher yield of maize.

If a farmer wants to quantify his benefits in terms of money considering prices of maize and pigeonpea at Rs. 4 and Rs. 6 per kg, respectively; the price equivalent ratio (PER) of each intercropping situation is to be determined and comparisons are to be made.

Crops	Yield (kg/ha)	Price (Rs/kg)	Total (Rs/kg)	Returns (Rs/kg)	PER
Sole maize	3398	4	13,592/-		
Sole Pigeon peas	1035	6	6,210/- 19802 -------- = 9901 2	(13,592+6210)/2	9901 ------ = 1.00 9901
1:1 Intercropped maize	2234	4	8,936/-		
1:1 Intercropped Pigeonpea	896	6	5,376/-	14,312	14312 ------ = 1.45 9901
2:1 Intercropped maize	3130	4	12,520/-		
2:1 Intercropped Pigeonpea	571	6	3,426/-	15,946/-	15946 ------ = 1.61 9901

Recommended spacing and plant population

1. Rice = 20 × 10 cm^2 plant population $= \frac{10000}{\times \; 20 \times 10} = \frac{100 \times 100}{1} \; \frac{\text{Area (in m}^2)}{\text{Spacing (m)}}$

   Since 1 ha rice plant population will be $= \frac{10000}{20 \times 10} \times \frac{100 \times 100}{1} = 5{,}00{,}000$ plants

2. Maize = 60×30 cm^2 $= \frac{10000}{60 \times 13} \times \frac{100 \times 100}{1} = 55{,}555$ plants

3. Sorghum = 45×15 cm^2 $= \frac{10000}{45 \times 15} \times \frac{100 \times 100}{1} = 1{,}48{,}148$ plants

4. Bajra = 50 × 10 cm^2 $= \frac{10000}{50 \times 10} \times \frac{100 \times 100}{1} = 2{,}00{,}000$ plants

5. Barley = 22.5×5 cm^2 $= \frac{10000}{22.5 \times 5} \times \frac{100 \times 100}{1} = 8{,}88{,}888$ plants

6. Wheat = 22.5×5 cm^2 $= \frac{10000}{22.5 \times} \times \frac{100 \times 100}{1} = 8{,}88{,}888$ plants

7. Gram = $30 \times 15\ cm^2 = 22.5 \times 5 = \frac{10000}{30 \times 10} \times \frac{100 \times 100}{1}$ = 2,22,222 plants

8. Mustard = $30 \times 15\ cm^2 = 22.5 \times 5 = \frac{10000}{30 \times 10} \times \frac{100 \times 100}{1}$ = 2,22,222 plants

9. Pea = $30 \times 10\ cm^2 = \frac{10000}{30 \times 101} \times \frac{100 \times 100}{}$ = 3,33,333 plants

10. Onion = $20 \times 10\ cm^2 = \frac{10000}{20 \times 10} \times \frac{100 \times 100}{1}$ = 5,00,000 plants

11. Cotton = $60 \times 30\ cm^2 = \frac{10000}{60 \times 30} \times \frac{100 \times 100}{1}$ = 55,555 plants

12. Safflower = $45 \times 15\ cm^2 = \frac{10000}{45 \times 15} \times \frac{100 \times 100}{1}$ = 1,48,148 plants

13. Castor = $90 \times 45\ cm^2 = \frac{10000}{90 \times 45} \times \frac{100 \times 100}{1}$ = 24,691 plants

14. Papaya = $2 \times 2\ m^2 = \frac{10000}{2}$ = 2,5000 plants

Rotational Intensity (RI)

This is calculated by counting the number of crops grown in a rotation and is multiplied by 100 and then divided by the duration of the rotation.

$$\text{Rotational intensity} = \frac{\text{No. of crops grown in rotation}}{\text{duration of the rotation (in yrs.)}} \times 100$$

This can be explained by the following example :

Maize – Early potato – Late potato – Mung – 1 year

Paddy – Early potato – Wheat – Urd   1 year

$$\text{Rotational intensity} = \frac{8}{2} \times 100 = 400\%$$

Agriculture year runs – 1 June to 31 May

1. Maize - Potato - Onion - RI = $\frac{3}{1} \times 100 = 300\%$
2. Rice – Tablepea – Onion – RI = 300%
3. Maize – Scane – Ratoon – 3 yrs, RI = 100%
4. Maize – Sugarcane – 2 yrs, RI = 100%
5. Paddy – Tablepea – Paddy – Chickpea : 2 yrs, RI = 200%
6. Maize – Early potato – Late potato – Chillis : 1 year, RI – 400%
7. Jowar + Guar – Berseem + Oats – 1 yr., RI = 200%
8. Jowar + Arhar – Fallow – wheat – 2 years, RI = 150%
9. Jowar + Arhar - Sunnhemp - wheat - 2yrs. RI = $\frac{300}{2} \times 100 = 150\%$
10. Sunnhemp - Sugarcane - 2 yrs, RI = $\frac{1 \times 100}{2} = 50\%$

    For green manuring only sunnhemp not taken into consideration as crop grown:
11. Maize (for cob) – E. potato – L. potato – sugarcane – Ratoon – Berseem.

    3 yrs, RI = $\frac{6 \times 100}{3} = 200\%$
12. Paddy - Pea - Scane - Rattoon - Lobia - 3 yrs., $\frac{5 \times 100}{3} = 166.66\%$
13. Fallow - Scane - Ratoo - 3 yrs, RI = $\frac{2}{3} \times 100 = 66.66\%$
14. Paddy – Field Pea – Scane – Ratoon – Lobia – wheat – 4 yrs.

    How does this rotation take four years.

	Crop	Sowing	Harvesting
1.	Paddy	July 2008	Oct. 2008
2.	Pea	Oct/Nov. 2008	Feb 2009
3.	S'cane	Feb. 2009	Feb. 2010
4.	Ratoon	Feb. 2010	July 2011
5.	Lobia	June/July 2011	Oct. 2011
6.	Wheat	Oct./Nov. 2011	April 2012

Here	July, 2008 - Feb, 2009 -	1 year
	Feb, 2009 - Feb, 2010 -	1 year
	Feb, 2010 - June, 2011 -	1 year
	June, 2011 - May, 2012 -	1 year
	Total	4 years

$$RI = \frac{6 \times 100}{4} = 150\%$$

Agriculture year runs from : 1 June to 31st May

1. Maize - Potato - Onion = 1 yr.

$$RI = \frac{3}{1} \times 100 = 300\%$$

2. Rice - Tablepea - Onion = 1 yr.

$$RI = \frac{3}{1} \times 100 = 300\%$$

3. Maize - Sugarcane - Ratoon = 3 yrs.

$$RI = \frac{3}{3} \times 100 = 300\%$$

4. Paddy - Pea - Paddy - Gram = 2 yrs.

$RI = \frac{4}{2} \times 100 = 200\%$ (Because pea and gram are grown in the same season)

5. Maize - Early potato - Late Potato - Chillies = 1 yr.

$$RI = \frac{4}{1} \times 100 = 400\%$$

6. Jowar + Guar potato - Berseem - Oats = 1 yrs.

$$RI = \frac{2}{1} \times 100 = 200\%$$

7. Jowar + Arhar - Wheat = 2 yrs.

   Here *jowar* and arhar are sown together but Arhar is taken as Rabi crop also. Therefore, rotation takes 2 yrs.

8. Sanai - Sugarcane : 2 yrs (Since, crop taken as green manuring is not considered as grown crop)

$$RI = \frac{1}{2} \times 100 = 50\%$$

9. Maize for cob – E. potato – L. potato – Sugarcane – Ratoon – Berseem

   $RI = \frac{6}{3} \times 100 = 200\%$

10. Paddy – Pea – Sugarcane – Ratoon – Lobia : 3 yrs.

    $RI = \frac{5}{3} \times 100 = 166.66\%$

11. Fallow – Sugarcane – Ratoon : 2 yrs.

    $RI = \frac{2}{2} \times 100 = 100\%$

12. Paddy – pea – Sugarcane – Ratoon – Wheat : 3 yrs.

    $RI = \frac{5}{3} \times 100 = 166.66\%$

13. Paddy – Pea – Sugarcane – Ratoon – Lobia – Wheat : 4 yrs.

    $RI = \frac{6}{4} \times 100 = 150\%$

14. Unirrigated crop rotation (Calculate RI)

Maize – Wheat	:	$= \frac{2}{1} \times 100 = 200\%$ (for Maize – Wheat to Mung – Wheat)
Jowar – Gram	:	
Maize – Barley	:	
Maize – Safflower	:	
Maize – Taramira	:	
Bajra – Wheat/Barley/Safflower	:	
Mung – Wheat	:	
Bajra – Fallow	:	$= \frac{1}{1} \times 100 = 100\%$

Irrigated crop rotation (Calculate RI)

a)	Paddy – Wheat – Mung	:	RI	$= \frac{3}{1} \times 100 = 300\%$ (for a–d)
b)	Maize – Potato – Wheat	:	RI	
c)	Maize – Wheat – Mung	:	RI	
d)	Maize – Mung – Safflower	:	RI	
e)	Jowar (Chari)- Barseem + Sarson – Bajra	:	RI	$= \frac{3}{1} \times 100 = 300\%$
f)	Paddy / Maize – Sugarcane = Rotton – Mung	:	RI	= 150%
g)	Maize = Cowpea – Barseem + Sarson – Bajra	:	RI	$= \frac{3}{1} \times 100 = 300\%$
h)	Sorghum + cowpea - Berseem + Mustard – Oats	:	RI	$= \frac{3}{1} \times 100 = 300\%$
i)	Maize – Potato – Wheat – Mung – Maize – Toria – Wheat – Mung	:	RI	$= \frac{8}{2} \times 100 = 400\%$

Cropping intensity (Cropping index)

Cropping intensity is the ratio of total cropped area to net cultivated area which is multiplied by 100 and represented in percentage.

$$CI = \frac{\text{Total cropped area (in a year)}}{\text{Net cultivated area}} \times 100$$

It does not consider the length of growing periods in crops.

Name of crops	Area (in ha)	Name of crops	Area (in ha)	Name of crops	Area (in ha)
Maize (green)	2.00	Paddy	1.00	-	1.00
Potato	2.50	Mustard and	1.50	Berseem	1.00
Sugarcane	4.00	Urd	2.00	Lobia	0.50
Mung	1.00	Cauliflower	1.00	Onion	1.00
Cucurbit	1.00	Maize (Fodder)	0.50		
Leafy vegetable	1.00				

Total net cultivated area = 9 (Net), Total Gross = 21 ha $CI = \frac{21}{9} \times 100 = 233.33\%$

Cropping scheme

The cropping scheme is a plan according to which crops are grown on individual plots of a farm during a given period of time with the object of a obtaining maximum return from each crop without impairing the soil fertility. Thus, a cropping scheme is related to the most profitable use of resources, land, labour, capital and management so that maximum net income may be obtained from the farm as a whole with proper restoration of soil fertility.

Principles of cropping (cropping scheme)

The following points are considered while framing a cropping scheme for a farm.

1. *Area under building and layout*

A proportionate area is allocated for buildings and layout before distributing the farm area under different plots of crops. The area under this item goes on decreasing as the total farm area increases and vice-versa. According to practices, if the farm area is 50 hectares or less then the area under buildings and layout allotted is 8-10 per cent of the total area, but if the farm area is more than 50 hectares then the area alloted is only 5 per cent.

2. *Number of plots*

The number of plots should be either equal to the duration of rotation or a multiple of the rotation. eg. if the total duration of rotations in a cropping scheme is 5 years, then the number of plots must be either five or ten, fifteen, twenty etc. so that each of the crop sown in the rotation could be grown in equal acreage.

3. *Selection of the crops*

Selection of crops depends upon the situation of the farm. Given below are the suitable crops which may be grown according to the farm situation or location for maximum profit.

a) *Near a city*. If the farm is near a city, the farmer should grow vegetables on at least 60 per cent of his net cultivated area. In the remaining area, he should grow fodder crops on at least 10-15 per cent of his net cultivated area.

b) *Near a sugar factory*. If the farm is situated near a sugar factory then the farmer should put at least 60 per cent of his net cultivated area under sugarcane or sugarbeet. On the remaining area, he sould grow vegetables and grain crop for local supply.

c) *Near a canning factory*. In such a locality, the farmer should grow vegetables eg. peas, tomato, etc., on at least 60 pcer cent of his net cultivated area. On the remaining area, he could have orchards.

d) *Near a Vanaspati factory*. The farmer should grow oilseed crops eg. groundnut, sunflower etc. in such a locality.

e) *Near a dairy farm*. If the farm is located near a dairy farm, the owner should go for fodder cultivation. Apart from fodder he can raise oilseeds and pulses on 10-15 per cent of his net cultivated area in order to supply concentrates to the dairy cattle along with fodder.

f) *Near a cold storage*. In such areas, the farmer should grow potatoes, onions, etc.

g) *Near a highway or railway station.* If the farm is situated near a railway station, port or highway, the farmer can go for the cultivation of perishable crops like vegetables, etc. In such cases, the farm is considered to be within the proximity of a city area.

4. *Facilities available on the farm*

Selection of crops also depends upon the facilities available for power, irrigation, inputs, labour, transport, etc. If these facilities are available, the farmer should adopt an intensive cropping scheme in order to maximise profits. Under an intensive cropping scheme, vegetables are grown because they fetch higher prices.

5. *Area under Individual plots*

The area of individual plots should be equal unless topography, soil conditions, or any other reason compel the farmer to layout his farm into uneven plot sizes.

6. *Area allotted to crops for calculation of cropping intensity*

If the farmer takes crop mixture then the relative area under different crops has to be considered, so that none of them area adversely affected due to their mutual competition. It is, therefore, essential that the ratio of main and sub crops are worked out properly before the crops are grown. For the purpose of calculation, both crops are considered to be one crop only.

7. *A crop planning which can utilize inputs and other resources available on the farm without wastage should be adopted by the farmer. Apart from this, the domestic needs are to be fulfilled.*

8. *The farmer should include at least one leguminous crop in a year in his crop rotation for proper maintenance of soil fertility and other physico-chemical properties.*

Suggestions of cropping schemes

(A) Suggest a suitable cropping scheme for a farm of 10 hectares which is situated near a sugar mill. The farmer has a tube well to irrigate his land. Calculate the cropping intensity and rotational intensity of the farm.

Solution : Under such cases we have to assume some of the things before we solve the problem.

Cropping Scheme

Plot No.	Area in ha.	1st Year *Kharif*	*Rabi*	*Zaid*	2nd year *Kharif*	*Rabi*	*Zaid*	3rd year *Kharif*	*Rabi*	*Zaid*
1.	1.0	Maize	Potato	Sugarcane	-	-	Ratoon			Mung
2.	1.0	_	Sugarcane	-	-	Ratoon	Mung	Maize	Potato	Sugar-cane ratoon
3.	1.0	Ratoon	-	Mung	Maize	Potato	Sugarcane	Paddy		
4.	1.0	Paddy	Lahi T_9-potato	Urd	-	-				
5.	1.0	M.P. Chari	Berseem + maize + lobia	-	Paddy	Tablepea	Onion	Paddy	Tablepea	Onion
6.	1.0	Maize	Sugarcane + potato sugarcane ratoon & mustard urd	-	Ratoon & mustard	-	Ratoon & mustard		Urd	
7.	1.0	-	Sugarcane	-	Maize	-	Urd	Maize	Sugarcane & potato ratoon & mustard	
8.	1.0	-	Ratoon + mustard + urd	-	-	Sugarcane & potato				
9.	1.0	Leafy vegetables	Cauliflower	Cucurbits + onion	-					
Total	9.0									

Assumptions

1. 10 per cent of the farm area would go under building and layout because the question says that the farm has a gross area of 10 heatares. In case of net area of 10 hectares this deduction would not have been made.
2. The farmer has all the required inputs for the cropping.
3. The farmer will grow sugarcane on 60 per cent of his net cultivated area.

Cropped area

Name of crops	Area in hectares	Name of crops	Area (in ha)	Name of crops	Area (in ha)
Maize (grain)	2.0	Paddy	1.00	M.P. Chari	1.00
Potato	2.50	mustard & lahi	1.50	berseem	1.00
Sugarcane	4.00				
Mung	1.00	Urd	2.00	lobia	0.50
Cucurbit	1.00	cauliflower	1.00	onioin	1.00
Leafy vegetable	1.00	maize (fodder)	0.5		
Total	9.00	Cropping intensity = $\frac{21}{9} \times 100 = 233.3\%$			

Rotational intensity of the farm

Rotat-ion No.	No.of crops	Duration of the rotation in years	Intensity in percentage	Examples
1.	5	3	166.7	5/3 × 100
2.	4	1	400.0	4/1 × 100
3.	3	1	300.0	3/1 × 100
4.	4	3	133.3	4/3 × 100
5.	4	1	400.0	4/1 × 100
Total	20	1	222.2	20/9 × 100

$$\text{Intensity} = \frac{\text{Total no. of crops}}{\text{Duration of the rotations}} \times 100$$

4. Crop rotations
   i) Maize – potato – sugarcane – ratoon – mung - 3 years
   ii) Paddy – lahi T_9 – potato – urd - 1 year
   iii) M.P. chari – berseem – maize & lobia - 1 year

iv) Maize - sugarcane & potato-ratoon & mustard-urd - 3 years

v) Cucurbits - cauliflower - onion - leafy vegetables - 1 year

5. Net cultivated area = 10-1= 9 hectares

6. Number of plots = 9 (because the total duration of all the rotations in 9 years)

(B) Suggest a suitable cropping scheme for a 10 hectare's farm which has all the facilities for intensive cropping. The farmer has to supply green fodder to a neighbouring dairy farm. Find out the cropping intensity.

Assumption

1. Ten per cent i.e. one hectare area is under building and layout. Thus the net cultivated area remains nine hectares.
2. The farmer has all the input facilities.
3. Crop rotatinos to be adopted :
   a) Maize-berseem-M.P. Chari-1 year
   b) Jowar-lucerne-maize & lobia-1 year
   c) Napier & guar-napier & berseem-napier 1 year
   d) Paddy-wheat-mung-1 year
   e) Bhindi-early potato-late potato-cucurbit-1 year.

Total number of plots area five

Cropping Scheme

Plot No.	Area in hectare	1980			
		Kharif	*Rabi*	*Zaid*	**Remarks**
1.	2.0	Maize	berseem	M.P. Chari	
2.	2.0	Jowar	lucerne	maize & lobia	
3.	2.0	Napier & guar	Napier & berseem	Napier	
4.	2.0	Paddy	wheat	mung	
5.	1.0	Bhindi	E. potato L. potato	cucurbit	
6.	9 hectares				

Cropped area

Maize – 3	guar – 1	lobia – 1
Berseem – 3	paddy – 2	napier – 2
M.P. Chari – 2	wheat – 2	potato – 2
Lucerne – 2	bhindi – 1	

Total cropped area is 26

$$\text{Cropping intensity} = \frac{26}{9} \times 100 = 288.8 \text{ per cent}$$

(C) Suggest a cropping scheme for a farm situated near city which has a net cultivated area of 10 hectares and an efficient sewage line for irrigation. The farm has all the required facilities.

Assumptions

1. There is no need to give any area under building and layout because the farm has 10 hectares of net cultivated area.
2. The farm has all the required inputs and transport facilities, for a high cropping intensity.
3. The farm is situated near a city and the farmer would go for vegetable production.
4. Crop rotations :
   a) Bhindi-pea-spinach-cucurbit
   b) Cucurbit-early cauliflower- tomato-bhindi
   c) Cucurbit-early cauliflower-cabbage-amaranthus
   d) Maize (cobs)-early potato-late potato-cucurbit.
   e) Jowar (chari)-early potato-wheat-mung
   f) Lobia-early potato-onion-spinach
   g) Jowar (chari)-berseem-M.P. Chari
5. Total number of plots would be 7 because the total duration of rotation is 7 years.

Cropping Scheme

Plot No.	Area in ha.	1980 *Kharif*	*Rabi*	*Zaid*
1.	2.0	Bhindi	pea-spinach	cucurbit
2.	2.0	Cucurbit	tomato-cauliflower	bhindi
3.	1.0	Cucurbit	E. cabbage-cauliflower	amaranthus
4.	2.0	Maize (cobs)	E. potato-L. potato	cucurbit
5.	1.0	Jowar (chari)	E. potato-wheat	mung
6.	1.0	Lobia	E. potato-onion	spinach
7.	1.0	Jowar (chari)	berwseem	M.P. Chari
8.	10			

Cropped area

Bhindi-4 (2+2)	maize-2	pea-2
Potato-6 (2+2+1+1)	spinach-3(2+1)	cucurbit-7 (2+2+1+1)
Cauliflower-3(2+1)	tomato-2	wheat-1
Amaranthus-1	jowar-2 (1+1)	wheat-1
Mung-1	lobia-1	onion-1
Berseem-1	M.P. Chari-1	
16 +	11 +	12=39

Total cropped area in 39 ha.

$$\text{Cropping intensity} = \frac{39}{10} \times 100 = 390 \text{ per cent}$$

(D) A village farmer wants to have a cropping scheme for his 10 hectare's farm. He has all the facilities required for crop production including irrigation at the farm. But there is neither a market near by nor does the farmer has transport facilities. Suggest cropping scheme for the farm and find out the cropping intensity.

Assumptions

1. There are no market or transport facilities, therefore, the farmer will grow cereals, pulses and oilseeds because these crops can be stored for a longer period and sold as and when needed.
2. The farmer has inputs and irrigation and hence he can go for high intensity cropping.

3. Ther farmer will grow vegetables for his domestic needs and fodder for his cattle, but the area under these crops would be low.
4. The farmer will grow vegetables for his domestic needs and fodder for his cattle, but the area under these crops would be low.
5. The farmer has to leave 1 hectare area for roads, channels, building etc. Thus the net cultivated area will be 9 hectares only.
6. Crop rotation : Calculate the R.I. of following
   a) Paddy-wheat-urd
   b) Maize-early potato-wheat-cheena
   c) Groundnut-wheat-sunflower & urd
   d) Paddy-mustard-mung
   e) Jowar (chari)-berseem-M.P. Chari
   f) Bhindi-early cauliflower-M.P. Chari

(E) Suggest a suitable cropping cropping scheme for a farm of 10 hectares, situated in Jhansi district which has no irrigation facilities. Find out the cropping intensity of the farm.

Assumption

1. The farmer will grow cereals, pulses and oilseeds which are drought resistant because the farm depends upon rain water only.

Cropping Scheme and Cropped Area

Plot No.	**Area in ha.**	**1988** *Kharif*	*Rabi*	**Cropped Area** **Crop**	**Area (in ha)**
1.	1.0	groundnut	wheat	groundnut	1.0
2.	1.0	castor	-	wheat	2.00
3.	1.5	jowar	gram	castor	1.0
4.	1.5	bajra	mustard	jowar	2.25
5.	1.5	arhar & jowar	-	gram	1.50
6.	1.5	til	linseed	bajra	1.50
7.	1.0	mung	wheat	mustard	1.50
8.	-	-	-	arhar	0.75
9.	-	-	-	til	1.50
10.	-	-	-	linseed	1.50
11.	-	-	-	mung	1.00
Total net area=9.0			Total		15.50

$$\text{Cropping intensity} = \frac{29.5}{9} \times 100 = 327.78 \text{ per cent}$$

Cropping Scheme and Cropped Area

Plot No.	Area of plot (in ha.)	1988			Cropped Area				
		Kharif	*Rabi*	*Zaid*	Crop	Area	Crop	Area	Total
1.	2.0	Paddy	wheat	Urd	paddy	4.00	jowar	0.5	4.5
2.	2.0	Maize	potato-wheat-chena	-	wheat	6.00	berseem	0.5	6.5
3.	2.0	Groundnut	whet	sunflower+urd	urd	3.00	M.P. Chari	0.5	3.5
4.	2.0	Paddy	mustard	Mung	maize	2.00	bhindi	0.5	2.5
5.	0.5	Jowar(chari)	berseem	M.P. Chari	potato	2.00	cauliflower	0.5	2.5
6.	0.5	Bhindi	e.cauliflower	Tomato	chena	2.00	tomato	0.5	2.5
				Cucurbit	groundnut	2.00	cucurbit	0.5	2.5
					sunflower	1.00			1.0
					mustard	2.00			2.0
					mung	2.00			2.0
Total net area = 9						26.00		3.5 = 29.5	

Example of cropping schemes

(A) Suggest a suitable cropping scheme for a farm of 10 ha which is suited near a sugarmill. The farmer has a tube well to irrigate his land. Calculate the cropping intensity and rotational intensity of the farm.

Solution: Under such cases we have to assume some of the things before we solve the problem.

Assumptions

i) 10% Farm area would go under building and layout.

ii) The former has all the required inputs for the cropping.

iii) The farmer will grow sugarcane on 60 per cent of his net cultivated area.

(B) Suggest a suitable cropping scheme for a 10 ha's farm which has all facilities for intensive cropping. The farmer has to supply green fodder to a neighbouring dairy farm. Find out the cropping intensity.

10% ± i.e. = 1 ha 10-1 = 9 ha

Cropping Scheme

Plot No.	Area (in ha.)	Kharif	Season Rabi	Zaid	Remarks	Cropped area
1.	2.0	Maize	Berseem	MP Chari	= 300%	6
2.	2.0	Jowar	Lucerne	Maize + Lobia	= 300%	6
3.	2.0	Napier + Guar	Napier + Berseem	Napier	= 300%	4
4.	2.0	Paddy	Wheat	Mung	= 300%	6
5.	1.0	Bhindi	E. Potato L. Potato	Cucurbit	= 400%	4
	9 ha.			$CI = \frac{26}{9} \times 100 = 288.89\%$		26 ha.

Note : Napier grass in case of season *kharif* and *rabi* is grown but yield of Napier is zero, although it occupy the area under above both seasons. Due to which there area is not added into final only intercrop *guar* & *berseem* is considered.

❑❑❑

Chapter-13

Agroforestry and Agronomy

Agroforestry is based on forestry, agriculture, animal husbandry and fisheries, land resource management, and other disciplines which all form the systematic background of land use system. It is a land use option that increases livelihood security and reduce vulnerability to climate and environmental change. It implies an awareness of interactions and feedback between man and environment, between demand and available resources in a given area, which, under certain conditions require optimization and sustained management rather than over exploitation. Agroforestry has two major components, agriculture and forestry, which form the "agroforestry system". That is why agroforestry, although not new in itself, requires new strategies and technologies as compared to traditional or modern agriculture and forestry.

Agroforestry has attracted the interest of scientists and development planners because it provides variety of products for meeting varied requirements of the people, insurance against risks caused by weather aberrations, controlling erosion hazards and ensuring sustainable production of the land on a long-term basis. It is a viable alternative to prevent and mitigate climate change and to meet global environment restoration commitments, agroforestry is an ideal option to increase productivity of wastelands, increase tree cover out side the forest, and reduce human pressure on forests under different agro-ecological regions.

Agronomy first appears in the early 1800s to describe "cultivation of the field," Today the definition is still much the same, though more descriptive in scope. Agronomy now refers to the agricultural subdiscipline that study all aspects of soil management crop production, and all the ways in which soils and crops interact. Soil management includes issues, *viz.*, soil classification, irrigation and drainage, soil fertility and tillage. Crop science generally refers to field crops like corn, wheat, soybean, rice, etc., and may study the selection, breeding, management and post-harvest treatment of crops, plant breeding, plant nutrition, and weed science. These soil management and cultural practices used for cultivation of crop also used by the agronomist when he evolves in doing agroforestry research/practices. Besides that in multiple cropping systems, agronomy also deals with resource use efficiency in respect to light, water and nutrients. The same principles also use in agroforestry with tree and crop/grass to increase resource use efficiency by using different agronomic practices. Sorghum + pigeonpea intercrop is the best example for higher resources use efficiency over sole crop studied at ICRISAT, Hyderabad. The sorghum + pigeonpea intercropping system has higher conversion efficiency of light early in growth and same intercrop also benefited from capturing more water.

This chapter examines the analogy of agronomy and agroforestry and importance of agroforestry in the current scenario. Synthesis of available literature also helps to identify the future direction of research and better understanding in between agronomy and agroforestry.

Analogy between agronomy and agroforestry

Agroforestry may be compared in concept with the multiple cropping systems of agriculture that have become increasingly popular over the last 20 years. Thus the, term multiple cropping will be the generic term comparable in many respects to the term agroforestry.

The term mixed cropping might be considered to be similar to that of agri-silviculture as in mixed cropping 2 or more crops are grown simultaneously with no distinct row arrangement on the same piece of land. The spatial arrangements of the crops in mixed cropping systems or in agri-silvicultural systems might be hapazard or might be organized. In these, a seeming multiplicity of agricultural species is grown in mixture. The species appear to have been chosen because their several root systems tap different layers of the soil for nutrients and water, because they possess different solar energy requirements, because together they simulate the many- storeyed physiognomy of natural tropical forests, and because in general the species complements rather compete with each other. Exact parallels of the hapazard mixed cropping systems; of agriculture are difficult to find in agri-silviculture, but the practice in north-east India and Kenya, for example, of having patches of annual crops interspersed with patches of forest tree crops, which together form forestry agri-silvicultural mosaics, might be considered to be similar in many respects to the mixed cropping systems of the agriculture.

The organized systems of mixed cropping in agriculture are normally referred to as intercropping or row-inter- cropping systems. In these, of course, the different species are grown in rows and the populations of each are fixed. This is similar to the traditional agri-silvicultural systems of the foresters, except, of course, that the forest trees are intercropped with agricultural species. It is suggested that the categories of agri-silviculture are not closed and besides to having single rows of trees inter-cropped with rows of agricultural crops it might be beneficial on some sites and for some trees and agricultural species, and under some socio-economic conditions, to have several rows or strips of trees that are intercropped with several rows of agricultural crops. The system might also be expanded to include the transfer of litter and/ or lopped branches from the strips of forest trees to the agricultural crops to provide woody mulch.

To extend the analogy between multiple cropping agricultural systems and agroforestry still further the "sequential cropping" systems of agriculture, in which more than one crop is grown on the same piece of land at different times of the year, might be likened to the planned fallows of agroforestry in which forest land is cleared and cultivated and then tree species chosen mainly due to their rapid soil ameliorating characteristics, are established when agricultural yields become uneconomic. These trees are in turn felled after the original soil fertility level is restored or improved, and the sequence is continued. In nutshell, "sequential cropping" in agroforestry is a system of planned fallow that is designed to control and harness shifting cultivation; and to reduce the fallow period without deterioration of the site.

The analogy between the multiple cropping systems of the agriculture and agroforestry systems may be pursued even further. But principles and concepts of agroforestry might not he wholly intelligible to agricultural scientists, and that the forester might learn a great deal from the research which has already been conducted on intercropping systems by agricultural scientists. When, we compare agri-silviculture with intercropping system, it is true that the basic difference between the two systems is that in agri-silvicultural systems in which we are attempting to grow trees with agricultural crops, while in intercropping systems the general practice is to grow different species of agricultural crops together in different rows. There are differences, therefore, between the systems, with respects to both spatial and temporal competition; for tree species not only possess different habits but they are also perennial. Nevertheless, the problem that is common to both the systems is that of the potentiality of competition among the different types of species for solar energy, for nutrients and for water. This problem is probably more difficult of solution in agri-silviculture than in intercropping systems. We must therefore, in agroforestry, address ourselves to the selection of species that are compatible with each other, and indeed complement to each other.

Agronomy contributions to agroforestry

Agronomic practices those are underutilized and new agronomic practices could be developed to increase productivity of agroforestry system under biotic and abiotic stresses. The research conducted on various aspects of agronomy and its contribution for resource conservation may be useful for the agroforesters. For example, in arid ecosystem, practice of using crop residues as stubble mulch is an effective measure of protection to the farm-land from wind erosion. Crop residues of 2 to 5 tonnes/ha and pearlmillet stubbles of 45 cm height were found very effective in preventing the blowing of sand from sandy soils. Similarly, vegetative barriers placed in the path of wind reduce its velocity near the ground by exerting a drag on the wind and deflecting the wind stream. The effectiveness of shelterbelts in reducing wind velocity depends on many factors *viz.*, wind velocity, direction, shape, width, tree height and density etc. Other agronomic practices other than windbreaks such as proper tillage, conservation and recycling of rainwater, use of sprinkler and drip irrigation methods are also helpful to increase production under such stresses.

Weather aberrations

It is always growing uncertainty mainly in agricultural and farm sector in India where the farming is mostly at the mercy of weather conditions. Approximately 30 to 60% reduction in potential crop yield is due to abiotic stresses and majority of it is attributed to drought.

Appreciating the contribution of abiotic stresses in Indian agriculture, ICAR recently established a research institute on abiotic stress at Baramati (Maharashtra) to undertake basic and strategic research. Broadly, major environmental constraints or the abiotic stresses are drought, heat (high temperature), low temperature/cold/frost, salinity and alkalinity, floods/hail- storms/high wind etc. Under such situations the role of agronomists is significant for making contingent plan/strategy to sort out the problems. The choice of an appropriate planting date and suitable crops/varieties may have a considerable effect on water use efficiency in arid and semi-arid regions. Advancing the sowing time, especially in post-monsoon season has been found to create moisture sufficiency for growth and nutrient uptake.

Drought

It is always associated with high temperature. Drought is classified as meteorological, hydrological and agricultural. The type or severity of abiotic stresses often varies with the agro-climatic locations and the crop. Intensity of drought is a ratio of actual evapotranspiration to potential evapotranspiration (AFT/PET) during the growing seasons at various phenophases of crop. The crops being the integral components of any given agroforestry system, the success of crop under the system will be one of the major determinants for developing a full fledge system. The shortage of rainfall and consequent drought conditions are the most serious constraints to crop production particularly in arid and semi-arid areas. The overall problem is very complex as the crops are subjected to wide range of soil moisture conditions and to diverse interplay of environmentl factors. Soil moisture variability caused mainly by rainfall variations may he of different nature, for example crop may be soil moisture deficit at any stage of growth due to erratic distribution of rains, early rain followed by prolonged dry period and soil moisture stress at critical crop growth stages.

Selection of suitable crops, varieties and cropping system

To mitigate adverse effect of weather aberrations, long-term strategic approaches are needed to efficiently conserve and utilize rainwatcr on the land. Selection of suitahlc crops/varieties and cropping system may be one of the strategies to mitigate the problem related to drought. The duration of traditional crops/varieties grown in rainfed areas is often longer than the effective growing season length. These crops/varieties usually experience moisture stress, mostly during grain filling period. Introduction of a relatively short duration cultivars found favour with the environment in rainfed areas. In intercropping system, short duration and long duration crop may be intercropped to make the best use of resources. The same principles are also used in agroforestry system in which trees as woody perennial and crop as annual grown together. Besides to efficient resource use, intercropping imparts stability to productivity and reduces the risk of crop failure. In traditional agriculture, farmers have developed several contingent crop production technologies, through their experience over the age, to cope with aberrant weather conditions. The scientists have refined several traditional management practices and have developed new ones to mitigate the drought effects to a certain extent. These techniques used for crop production in agroforestry, tree also share the benefits. For instance, inter-culture, weed control, tillage, mulching etc. applied to the crop also benefits the trees in agroforestry system.

Efficient rainwater management

Efficient rainwater management acts as insurance for the crop during rain deficit periods. Management techniques that increase infiltration and soil water storage and decrease water losses by runoff, evaporation and evapotranspiration by weeds increase the amount of water retained in the soil for subsequent use by crop. Mechanical structure like contour or graded bunding and bench terracing are the pre-requisites for water conservation in sloppy lands. Practices such as profile modification, deep tillage, mulching/vertical mulching, dead furrows, compartment bunding and graded border strips can increase infiltration and reduce runoff. The National Research Centre for Agroforestry, Jhansi has developed planting techniques of aonla for *in-situ* moisture conservation to mitigate drought impact on establishment and fruiting of aonla tree. The technique involves digging of pits (1m × 1m × 1m) before rain and later refilled with 80:20 ratio of FYM and soil mixture up to 0.60m depth. One-year-old budded plant of the improved variety of aonla can be planted after onset of monsoon. The small pieces of stones locally available in field are collected and placed on the refilled mixture in layer by layer up to 0.90m depth. The rest of 0.10m depth of the pits is kept as open. Small ridge is prepared around the pit and an inlet is made on upper side of the slope to collect runoff water during rain (Fig. 13.1 and 13.2).

In this process the compost acts as a sponge for retaining the moisture and small stones impede evaporation. This technique may harvest approximately 200 litres additional water during excess rains provided 50 % porosity is available in the layer of stone mulching otherwise this precious amount of water is lost through runoff where *in-situ* moisture conservation technique are not being used.

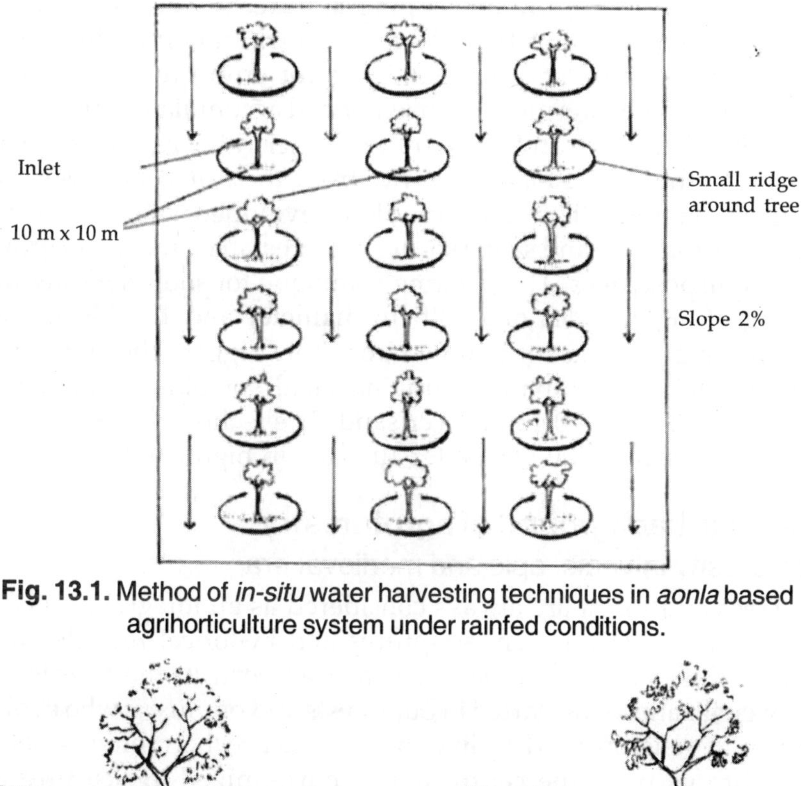

Fig. 13.1. Method of *in-situ* water harvesting techniques in *aonla* based agrihorticulture system under rainfed conditions.

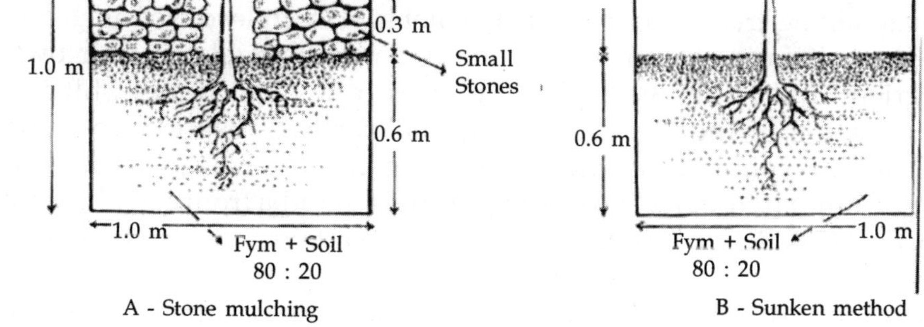

Fig. 13.2. Planting Techniques

Degraded salt-affected soils

Agricultural use of salt affected lands and water resources increases with increasing demands of food and fodder. Sustainable use of these lands for food and feed production will become a more serious issue. Removal of salts from the soil surface is neither possible nor practical; therefore attempts have been made to minimize adverse effect of salts on crop by developing agro-techniques. Central Soil Salinity Research Institute. Karnal (Haryana) has developed appropriate package of agro-techniques for crop production in salt affected soils. The salts move with water and accumulate in the surface soil or at top of ridge and centre of the wider ridge results in greater damage to the crops. Planting away from the salt zone lessens the extent of damage. The institute also developed special planting techniques for sodic and saline soil for better establishment and growth of multipurpose trees. The planting technique for sodic soils involves digging of auger holes of 20-30 cm diameter and 100-140cm deep with the help of a tractor mounted post-hole digger. The auger holes are filled back with a mixture of original alkali soil plus 3-4 kg gypsum plus 8-10 kg FYM + 10-15 kg river sand. The technique ensures more than 80 % tree survival even after 10 years in highly alkali soil.

Historical background of agroforestry

Agroforestty in vedic, epic and medieval era

Trees and forest are always considered as an integral part of the Indian culture. The ancient scriptures and historical records amply support this. The best Indian culture was born in the forests. The Aryan civilization was started in our forests and our *Rishis* who evolved the Hindu religion used to live in the forest in harmonious nature. The ashranis were the centre which harmonized agriculture and pasture with trees, animals and birds. The destruction of forest and cutting of trees created famine conditions, where planting and maintaining trees were regarded as noble acts. Indeed, so much has been written in our ancient literature that individuals on their own agricultural fields were planting trees in ancient times. Shifting cultivation in India is pre-historic and partly a response to agro-ecological condition in the region. Horticulture as coexistent with agriculture is found to have been prevalent in India from early historic period (500 BC to 1st century AD) when a certain amount of share in garden crops started to have enjoyed by the king for providing irrigation. A number of references are available in different texts of the Vedic literature.

Agroforestry during past 5 centuries

During the last 3 to 5 centuries, spectacular improvements have been made in tree plantation. The most important agroforestry practice is known from the *Kangeyam* tract of Tamil Nadu, where *Acacia leucophloea* + *Cenchrus setigerus* were perfected in silvipasture system. Similarly, in ravines of *Yamuna* and *Chambal*, trees, shrubs and bamboos with grasses were planted for rearing milk-producing '*Jamunapari*' breeds of goats and sheep. Scattered trees with *khejri* or *mehndi* in association with *bajra, jowar* and *chillies* were grown in the semi-arid area of Tamil Nadu. Tree plantation continued as demarcation and control against wind erosion throughout the country. Plantation of *khejri* trees for various uses on- farm was a common practice in Rajasthan. In the northeast, large cardamom (*Ammomum subalatum*) was introduced in Alder (*Alnus nepalensis*) forest in Sikkim Himalayas by the farmers more than 200 years ago, and now the system is being practised in more than 48,000 ha area in the north-eastern states.

In coastal area *Casuarina equisetjfilia* and other trees were grown in association with crops on farm lands for cash and to generate small timber. Live hedges were common as an agroforestry practice in which *Mehndi, Agave sisalana* and *Euphorbia species* were common. In paddy-growing areas *Pangamia glabra* and *Sesbania grandiflora* were grown, lopped annually and their leaves applied to fields as green manure. Application of green manure to paddy fields was common in Madhya Pradesh and Uttar Pradesh. In Western Ghats *Terminalia* leaves were harvested, spread on land, burnt and then *paddy, ragi* and *millets* were sown. Multistorey homesteads or home-gardens were in existence in Kerela, Karnataka, Tamil Nadu, Tripura, Assam and other north-eastern states as an important agroforestry practice. This practice is still followed in these states.

Agroforestry definition

There are several definitions of agroforestry given by the experts and scientists working on agroforestry. Looking to the available definition of agroforestry, there is a consensus that agroforestry is a land use system which involves trees with agricultural crop/grass and/or animals simultaneously or sequentially. The most commonly used definition is "Agroforestry is a collective name for land use system in which woody perennials (tree, shrubs etc.) are grown in association with herhaceous plants (crops, pastures) or livestock, in spatial arrangement, a rotation or both; there are usually both ecological and economic interactions between the trees and other components of the system".

Components of agroforestry

In general there are 4 or 5 basic sets of components which are managed by human beings in all agroforestry systems. Structurally, the system can be grouped as:

- Agri-silviculture system
- Agri-horticulture system
- Silvipastoral system
- Agri-silvipastoral system

On the basis of nature of components the following are common agroforestry systems found in different agroecological regions of India:

- Agrisilviculture (trees + crops)
- Boundary plantation (tree on boundary + crops)
- Block plantation (block of tree + block of crops)
- Energy plantation (trees + crops during initial years)
- Alley cropping (hedges ÷ crops)
- Agri-horticulture (fruit trees + crops)
- Agri-silvihorticulture (trees + fruit trees + crops)
- Agri-silvipasture (trees + crops + pasture/animals)
- Silvi-olericulture (tree + vegetables)
- Horti-pasture (fruit trees + pasture or animals)
- Horti-olericulture (fruit tree + vegetables)
- Silvi-pasture (trees + pasture/animals)
- Forage forestry (forage trees + pasture)
- Shelter-belts (trees + crops)
- Wind-breaks (trees + crops)
- Live fence (shrubs and under- trees on boundary)
- Silvi-horti-sericulture (trees/fruit trees + sericulture)
- Horti-apiculture (fruit trees + honeybee)
- Aqua-forestry (trees + fishes)

  Homestead (multiple combinations of trees, fruit trees, vegetable etc).

Interactions in agroforestry

Physiologists and ecologists have long argued that a sound understanding of the processes and mechanisms involved in resource capture and use and their interaction with the environment is essential for the development of more reliable and productive systems. There is still considerable doubt concerning the impact of such understanding after that impressive accumulation of knowledge of intercropping in recent years. The purpose of the discussion on the principles of resource capture, drawing heavily on the intercropping literature is to illustrate the difficulties involved. Appropriate methods of measurement and analysis are outlined and their application to agroforestry research is considered. Special attention is devoted to consideration of the key differences between intercropping and agroforestry systems, major opportunities for translating an improved understanding of these systems into increased productivity or improved sustainability, and the main bottlenecks' to future progress in research.

Competition between species in mixed stands (intraspecific competition) differs from that between plants within monocultures (interspecific competition) in that the component species of intercrops may impose different demands on the available resources. The intensity of competition is greatest when site requirements are similar, to the point where species with over-lapping niches may be unable to coexist within the same community. Vandermeer (1989) suggested that competition may he more severe between similar species with similar growth habits than that with species with contrasting growth habits. However, the opportunity for complementarity of resource use between species is restricted by the fact that all plants are competing for the same, usually finite, resource (light, CO_2, water, nutrients). Thus, there is extensive overlap between species in their resource requirements, with the partial exception of legumes which are capable of fixing atmospheric nitrogen.

The biological potentials and restraints of agroforestry systems and environmental responses of plant components within them, is tree-crop interface. The components of intercrops or agroforestry systems often differ greatly in size, with the result that the growth of the smaller understorey species may be inhibited by shading, and possibly also by competition for water and nutrients. Competition fur light is the primary limitation when water and nutrients are freely available. However, in many tropical systems, water (*e.g.* semi-arid regions) or nutrients availability (*e.g.* acidic, leached or degraded soils) rather than light is the major limiting factors. It is not always straightforward to establish which the primary limitation is when more than one factor is marginal. For instance, a species which establishes an early advantage in light capture through more rapid initial shoot growth may also exhibit greater root growth and hence resource capture because of the increased availability of photosynthate. This may in turn further improve shoot growth and light interception to the detriment of the less competitive species in the mixture. A better understanding of the root growth pattern of agroforestry trees is required for quantitative explanation of the competition process such as competition for water and nutrients, nutrient recycling etc. Though there is substantial information available on root studies for seasonal crops and up to some extent for forest species but only limited information is available on agroforestry situation.

In agroforestry, complementary or competitive effects depends upon (i) age and size of the trees, (ii) nature of the tree species. (iii) nature of the agricultural crops, (iv) availability of water, nutrients and light etc. In an inter- cropping system involving a legume and non-legume, part of the nitrogen fixed in the root nodule of the legume may become available to the non-legume component which is an example of complementarity. The compleinentarity in resource capture under mixed cropping/ intercropping system may he spatial or temporal. In temporal complementarity, the leaf canopies of component crops may occupy different vertical layers with taller component tolerant of strong light and high evaporative demand and shorter component favoring shade and high relative humidity. Multistoreyed cropping in coconut and planting shade trees in cocoa and tea plantation uses this principle. In similar manner under agroforestry system, exploitation of different layers of soil by the root systems of trees and crops may lead better utilization of resources with much less competition. In temporal complementarity, the yield advantages provided by the mixture cannot always he explained by more effective use of growth resources at specific times. Indeed, there are substantial opportunities for temporal complementarity if species make their major demands on available resources at different times, thereby reducing the possibility of competition. For example, the usually advantageous intercropping of the slow-growing pigeonpea with fast-growing cereals *viz.*, maize or sorghum produces only limited increase in total light or water use hut nevertheless provides intercrop advantages of 40 to 70%.

Benefits of agroforestry

Enhanced productivity

There are ample evidences to show that the overall (biomass) productivity, soil fertility improvement, soil conservation, nutrient cycling, micro-climate improvement, carbon sequestration potential of an agroforestry system is generally greater than an annual system although not necessarily greater than a forestry or grassland system. Some evidences on above mentioned benefits under various agroforestry systems are described in following subheads and different agroforestry systems are illustrated in Fig. 13.3 (a-f).

In agroforestry, the potentially higher productivity could he due to the capture of more growth resources *e.g.* light or water or due to improved soil fertility. Competition, which is a negative effect in this context, is usually a significant factor in tree-crop system; even then there is evidence of increased combined productivity by both components. If we take the example of poplar (*Populus deltoides*), which is a popular species in agroforestry system in western Uttar Pradesh, Uttarakhand, Haryana and Punjab. *Poplar* being a deciduous in nature it allows agricultural crops to grow under it without adversely affecting the yield. The studies on agri-silviculture with various intercrops showed that upto first 2 years the returns obtained from sugarcane + *poplar* combination are higher than any other crop, as the preparatory and cultural operation for both of these are complementary. However, third year onwards shade-loving crop like turmeric and ginger can be grown under it. During the winter (*rabi*) wheat can be grown till its harvest. In another study, economics of *poplar* was estimated by Singh and Dhaliwal (2005) under different agroforestry system *viz.* AFS-I (block plantation of *poplar* with intercropping wheat-*kharif* fodder, AFS-II (block plantation of *poplar* with intercropping sugarcane for first 3-years and then wheat-*kharif* fodder for the successive years), AFS-III (block plantation of poplar with intercropping potato-*kharif* fodder for first year and then wheat-*kharif* fodder for the successive years) at two sites (Ludhiana and Ropar district of Punjab). These agroforestry systems were compared with paddy-wheat pure cropping system (without yield reduction). The net present values (NPV) were calculated at 12.5% discounted rate. The highest NPV was Rs 2,86,687/ha in Ludhiana for AFS-II while it was Rs 3,13,420/ha for AFS-II in Ropar district. The B:C ratio was highest (4.79) for AFS-II in Ludhiana and 5.43 for AFS-II in Ropar district. The agroforestry systems gave higher economic return than paddy-wheat pure cropping (Fig. 13.4). Chauhan (2000) estimated the economic of aromatic crops like lemongrass (*Cymbopogon flexuosus*), citronella java (*C. winterianus*), palmarosa (*C. martinii*) and Japanese mint (*Mentha arvensis*) with poplar at Pantnagar. The cost of cultivation varied from Rs 8,520 to 21,625/ha/year with different crops. The net income obtained from *poplar* with aromatic crops was higher than growing of sole *poplar* or sole crop. It indicates that agroforestry is profitable venture than pure cropping (Table 13.1).

Several studies in different parts of the country suggested that agroforestry is more profitable to farmers than agriculture or forestry for a particular area of land. Plantation of kinnow at 4m × 4m spacing and Bhabar (*Eulaliopsis hinata*) at 50cm × 50cm after minor leveling at Relmajra provided early return to the farmers. The system was highly profitable and quickly paid off the leveling cost. In another study, a silvipastoral system involving Bhahar in association with *Bauhinia purpurea* or *Grewia optiva* appear to be the most promising for sustained biomass production from the degraded houldery riverbed lands of Doon valley.

National Research Centre for Agroforestry has been working since 1989 on agri-silvi-horticulture system which included 4 varieties of *aonla viz.*, Chakaiya, Kanchan, Krishna and NA 7 as fruit trees, *Leucaena* as multipurpose tree and blackgram as intercrop in rainfed areas. The *Leucaena* was planted on both sides of the fruit trees at 2m distance. The aonla was planted at 10m × 6m and 5m × 6m spacing but 10m × 6m spacing was proved an ideal spacing among these and it was considered for calculating the yield and economics of the system. The cost of cultivation in first year which includes planting of fruit trees and *Leucaena* as multipurpose trees and cultivation of crop was Rs 8,666/ha but during next year the cost of cultivation was reduced and it was again increased with subsequent increase in the cost of input during different years. The gross income from the system was less during initial year but when fruiting started in aonla, the gross income was increased and it went up to Rs 60,712/ha at age of 13 years. Similarly the net income was positive in all the years except at first, third and six years when aonla had no fruiting/or less fruiting (Fig. 13.5). In age of 13 years, the B: Ç ratio from the system was 3.28 and on discounted rate it was 2.61 which indicated that aonla based agroforestry system is a profitable enterprise in marginal lands under rainfed conditions.

Fig. 13.3. Agroforestry system for food production (A) pulse crop (Vigna radiata) with Emblica officinatis (B) mustard crop (Brassica juncea with Dalbergia sissoo (C) wheat crop (Triticum aestivum) with Albizia procera (D) wheat crop with Encalyptus tereticomis; Agroforestry system for animal production (E) goat and sheep grazing under silvi-pastoral system; Agroforestry for food and bio-diesel and (F) whea crop with Jatropha curcas.

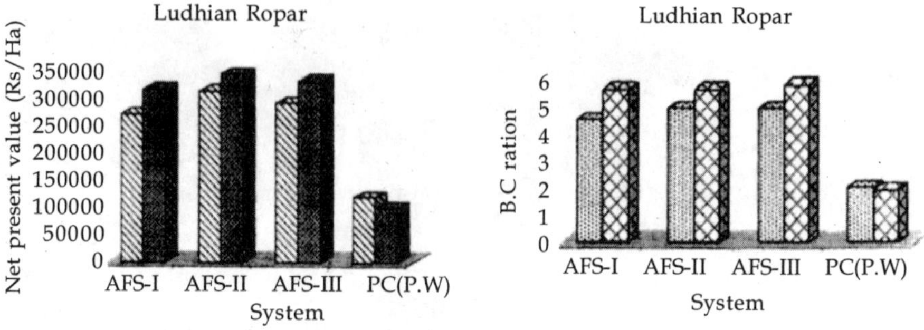

Note: AFS-I, Agroforestry system (Block plantation of popler with intercropping wheat-*kharif* fodder); AFS-II Agrotorestry system(Block plantation of poplar with intercropping sugarcane for first three year and wheat-kharif fodder for the successive years; AFS-III, Agroforestry system (Potato-*kharif* fodder for first year and then wheat-*kharif* fodder for successive years) and PC(P-W), purecropping (Paddy-wheat)

Fig. 13.4. Economics of poplar block plantation (with intercropping) and pure cropping at 12.5% discount in 6th year.

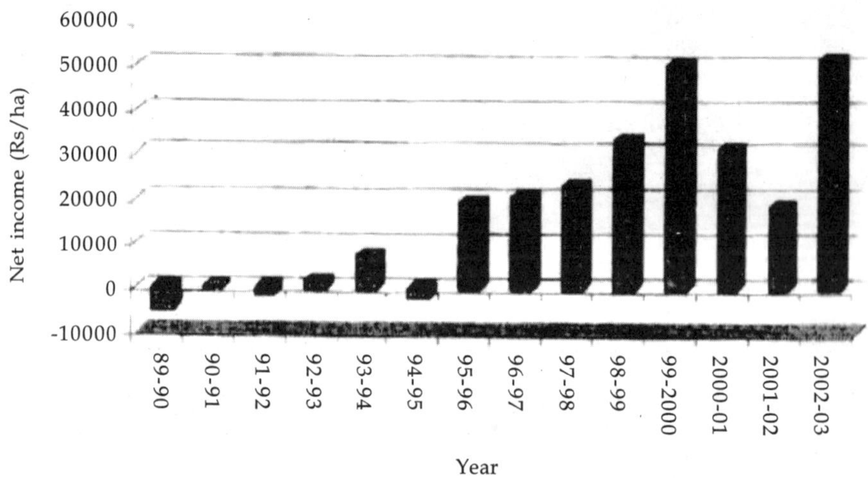

Fig. 13.5. Income from aonla based agroforestry during different years

Agroforestry and weed control

Weed control is one of the important biotic challenge for the agronomists. It not only affects crop yields but also acts as source of adulteration. Agroforestry has been viewed as an example of tree species competition in which a secondary species (*e.g.* tree species) controls the weeds and releases the main crop from the strong competition that may be offered by the weed. Most weeds are strong light demanding species and an effective way of controlling their incidence is through canopy closure and shading the undergrowth as in intercropping. Lower weed yield under hedge rows with *Acacia albida*, control of *Imperata* infestation in alley cropping and shift in weed composition following alley cropping with hedgerow species are some of the examples of reduced weed infestation by introducing agroforestry systems. In north-eastern India, the hot humid climate favours luxuriant weed growth that results in reduction of crop yields substantially and proper weed control can increase crop yields by 38% found that weed infestation could be greatly reduced under suitable agroforestry systems in the north-eastern India.

Enhancing soil fertility

The primary objective of soil conservation is to improve/maintain soil fertility. To achieve this, control of erosion, maintenance of organic matter and physical properties, organic matter addition, maintenance of nutrient is necessary. In this way agroforestry system constitute sustainable land use and helps to improve soils in a number of ways. Some of these beneficial affects are apparent in experiments carried out in different parts of the country. Tree crowns protect the soil from the impact of raindrops and create a specific micro-climate in the particular area. A comparison of different land use system (Table 13.2) gives a very clear picture about agroforestry system that increased soil organic carbon and available nutrients than growing sole tree or sole crop. Increase in amount of nitrogen and phosphorus and other macro-and micro-nutrients are reported in 14 -year old plantation of *khejri* (*Prosopis cineraria*) as compared to a bare site of *Prosopis juliflora* of the same age. The forage species produce higher biomass under *khejri* tree canopy duc to a high fertility status. There are evidences that soil under the trees were richer in organic matter and nutrients than those in the surrounds without tree cover. For *Acacia albida,* 50 to 100% increase in organic matter and nitrogen under canopy are known, together with increased water holding capacity. Alley cropping is very sound technology for improving the fertility of degraded low- activity clay soil in the tropics and the key to its success lies in the use of pruning as green manure or mulch.

Table 13.1. Economics (Rs/ha) of poplar intercropping with aromatic crops

Agroforestry system	Net return from intercrop	Net return from tree	Total net return in 5-year rotation	Average net return (Rs/ha/yr)
Sole *poplar*	-	1,28,450	1,28,450	25,690
Poplar + Lemon grass	82,950	1,35,000	2,17,950	43,590
Poplar + Citronella java	64,250	1,36,550	2,00,800	40,160
Poplar + Palmarosa	63,900	1,34,450	1,98,350	39,670
Poplar + Japanese mint	44,825	1,37,025	1,81,850	36,370
CD (P=0.05)	1,426	2,711	12,563	1,250

Table 13.2. Comparison of soil fertility improvement in sole tree, sole crop and agroforestry system

Treatment	Organic carbon (%)	Available nutrients (kg/ha)		
		N	P	K
Sole crop	0.48	252.5	20.2	280.5
Subabul alone	0.46	180.2	12.5	250.0
Shisham alone	0.40	192.5	12.5	260.7
Shisham + crop	0.61	265.3	23.5	260.7
Subabul + crop	0.72	283.5	25.5	320.5

Source: AICRPAF Annual Report, 1999-2000.

An experimental evidence under agri-silvi-horticultural system (*aonla* + *Leucaena* + blackgram) in rainfed conditions showed that after 9-years 65 to 109.4% organic carbon was increased under below canopy and 28.1 to 62.5% under open canopy as compared to initial value (0.32%). The % organic carbon under the canopy of aonla was higher under than in open canopy owing to complete falling of aonla leaves mostly limited to below tree canopy. An increase in organic carbon, available N, P and K content in khejri based silvipastoral system over no-khejri soil, advocating retention/plantation of khejri tree in pasture land to get higher fodder production and to meet requirement of food, fodder, fuel and small timber (Table 13.3).

Table 13.3. Imporatnt chemical characteristics of soil under sole Dhaman grass (*Cenchrus ciliaris*) and beneath *khejri* plantation in silvi-pastoral system.

Characteristics	Sole grass	Beneath *khejri*	In between *khejri* rows
N (kg/ha)	125.30	173.30	157.00
P (kg/yha)	19.50	29.90	22.40
K (kg/ha)	187.00	231.00	220.00
Organic carbon (%)	0.22	0.27	0.24
EC (dS/m)	0.05	0.09	0.05
pH	8.15	7.81	7.87

Source: Jaimini *et al.*, (2006)

An increase in soil organic carbon status of surface soil 0.39% to 0.52% under *Acacia nilotica* + *Saccharum munja* and 0.44% to 0.55% under *Acacia nilotica* + *Eulaliopsis binata* after 5 years and suggested that *Acacia* n/lot/ca + *Eulaliopsis hinata* are conservative but more productive and less competitive with trees and suitable for eco-friendly conservation and rehahilitation of degraded lands of Shivalik foot hills of sub-tropical northern India. Rehabilitation of degraded forests is possible through afforestation by adopting integrated land use planning with soil and water conservation measures on watershed basis observed that in agri-silviculture growing of *Albizia procera* with different pruning regimcs, the organic carbon of the soil increased by 13-16% form their initial values under different pruning regimes which was 5 to 6 times higher than growing of either sole tree or sole crop (Fig. 13.6).

In depth evaluation of soil chemical properties of traditional agroforestry system in north-eastern region indicated a spectacular increase in soil pH, organic C, exchangeable Ca, Mg, K, and build up of available P (Bray's P_2-P) under different agroforestry practices (AFP) within 10-15 years of practice. Accumulation of 2.91% organic C was observed under arecanut + jackfruit + black pepper + Cinnamom (tejpatra) followed by 1.85% under arecanut + betelvine + miscellaneous trees as against 0.78% only in a degraded land within 10-15 years of this practice. A sharp increase in exchangeable Ca, Mg, K and Na was observed in all the agroforestry interventions over adjoining degraded lands. The exchangcable Al, potential cause of infertility of these lands disappeared completely within 10-15 years of agroforestry practice. This was attributed to the addition of fresh organic matter which complexed exchangeable Al during decomposition possibly due to formation of Al-humate and accumulation of Ca. Mg. K and Na cations. It eventually increased soil pH by 0.6 to 1.7 units under these AFP. Thus, the potential AFP was Ibund to have built in dynamism for the restoration of soil fertility and sustained yield. Similar results were obtained when multipurpose trees were evaluated in an extremely P-deficient acid Alfisol in Meghalaya.

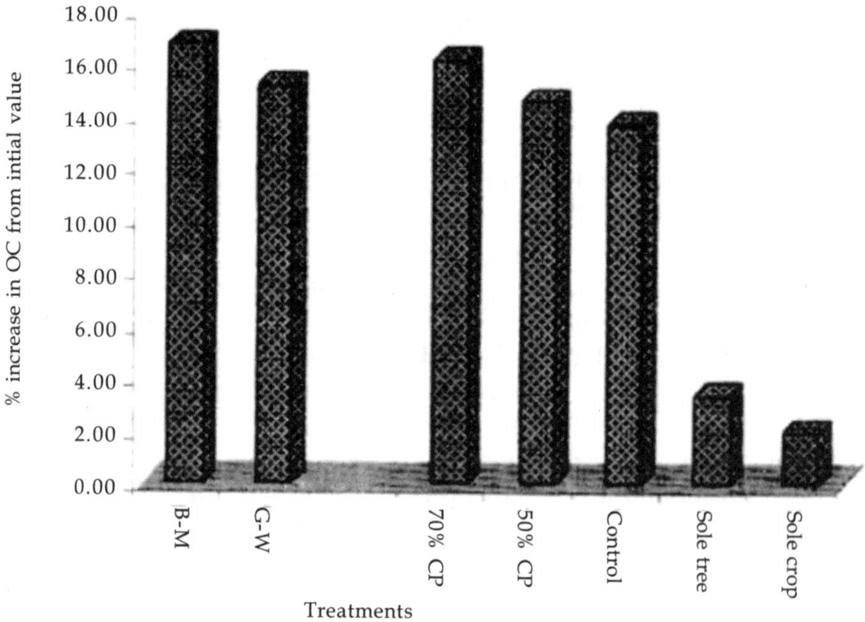

Note: B-M, Blackgram-mustard; G-W, Greengram-wheat; CP, Canoy pruning

Fig. 13.6. Increase in soil organic carbon under agri-silviculture and pure tree and pure crop from initial value after 3 years.

Soil conservation

Agroforestry systems on arabic lands envisage growing of trees and woody perennials on terrace risers, terrace edges, field bunds, as intercrops and as alley cropping in the shape of hedgerow plantation. Integrating trees on the fields act as natural sump for nutrients from deeper layers of soil, add biofertilizer, conserve moisture and enhance productivity of the system. The alley cropping with leguminous trees *viz.*, subabul (*Leucaena leucocephala*) has been most widely used on field bunds for producing mulch material for moisture conservation and nutrient recycling. Alley cropping with *Leucaena leucocephala* was effective for erosion control on sloping lands up to 30%. The contour-paired rows of *Leucaena hedge, Leucaena* and *Eucalyptus* trees, and 0.75m wide grass barrier at 1.0m interval in maize brought down runoff from 40 to 30% of rainfall and soil loss from 21 to 8 t/ha on 4% sloping land. Reduction in crop yield could be minimized by shifting the management of trees to contour hedgerows. The sediment deposition along the hedge and tree rows increased considerably with consequent reduction in soil loss. Application of *Leucaena* before or at harvesting of maize increased moisture conservation and yield of following wheat crop under Doon valley conditions.

In agroforestry, multi-strata practices simulate the protective nature of natural forest. Purposefully planted cover can take several forms that can differ in their ability to counter soil and water loss. Vegetative cover is the key to arrest runoff and soil loss. It has been estimated that soil erosion is accelerated to more than 900-fold when tree plantation were clean weeded or the litter was burnt. In a semi-arid Allisols degraded land established pastures reduced runoff to 26% and soil loss to 8%. In another study agroforestry system recorded runoff 2.7%, soil loss 0.54 tJhalyear, N loss 2.1 kg/haiyearand K loss 0.7 kg/ha/year with mean monsoon season rainfall of 899mm compared with runoff 25.4%, soil loss 3.21 tlha/ year, N loss 12.5 kg/ha/year and K loss 4.3 kg/ha/year from the field crops raised under same condition. On slopy lands contour hedgerows have consistently been demonstrated to be highly effective in controlling soil erosion even in as little as 18 months. The woody hedgerows provide a semi-permeable barrier to surface movement of water, while mulch from the trees reduces the impact of raindrops on the soil and minimizes splash and sheet erosion. By alternate land use systems like silvipasture, agri-silviculture and agri-horticulture, the productive and protective benefits from watershed management (WSM) are considerably higher than the investment, benefit: cost ratios ranging from I .92: I to 7.1:1 with 48 to 99 % reduction in runoff and 81 to 98% soil loss besides the check of out migration of population from 26.6% before the implementation of WSM programme to 9.3% during the project period.

Agroforestry systems have been developed using local resources and conservation-based measures in the northeastern hills (NEH) region. Suitable alternate land use systems involving agriculture, horticulture, forestry and agroforestry have been designed with the support of local natural resources for almost identical hydrological behaviour as under the natural system. The model land use suggests utilizing slopes below 50% towards lower foothills and valley lands for agricultural crops and pisciculture, middle slopes between 50 and 100% for horticulture and top slopes over 100% for forestry/silvipastoral establishment. Under agri-horti-silvi-pastoral systems, the reduction in runoff was 99% and in soil ios 98%. Combining fine-root system of grasses and legumes, such as *Stylosanthes guyanensis, Panicum maximum, Setaria* etc. and deep-root system of fodder trees, such as alder (*Alnus nepalensis*) in a silvi-pastoral system stabilizes terrace risers and provides multiple outputs. Silvipastoral system comprising *Alnus nepalensis*, pineapple and forage crops like *Panicum maximum* or *Setaria sphacelata* coupled with *S. guyanensis* in 1:1 ratio was found to be a sustainable agroforestry practice in soils having 30-60% slope. Forage yield of 13.5 tlha was obtained from the combination of *Stylo* and *Setaria*. In addition, 2.3 t/ha litter from *Alnus nepalensis* and 4,000 fruits/ha from pineapple were obtained. This system also restored the fertility of these soils.

Micro-climate improvement

The use ol trees as shelterbelts in areas that experience high wind or sand movement is well-established example of micro-climate improvement that resulted in improved yields. Increased agricultural production due to windbreaks and shelterbelts in India has heen reported by Rao and Sita Ram (1980). Establishment of micro-shelterbelts in arabic lands, by planting tall and fast-growing plant species *viz.*, castor bean on the windward side, and shorter crop such as vegetables in the leeward side of tall plants helped to increase the yield of lady's finger by 41% and of cowpea by 21% over the control. In spite of good results the community shelterbelts in arable lands are not very popular with farmers, especially as the technique normally cut across their field holding boundaries.

Experimental shelterbelts of *Acacia nilotica* and *Dalbergia sissoo* were established over 120km in square blocks at the Central Mechanized Farm of Central Arid Zone Research Institute at Suratgarh in Bikaner District, Rajasthan. Reduction in wind speed was greater during monsoon (July-September) than before the monsoon period (April-June). There was a 5-14% reduction in evaporation from April to July. In general, the use of shelterbelts brought about a 50% reduction in the magnitude of wind erosion. In the *Cassia siamea* shelterbelt, the loss was 184.3 kg/ha while in hare soil (i.e. without any shelterhelt) it was 346.8 kg/ha. The evidence for the beneficial effects of shade trees depends on the nature of the understorey crops. The clear effect is reported for crops that require shading for normal growth, e.g. black pepper. turmeric, cacao. In studies carried out in an agroforestry system, *Acacia tortilis* (7-year-old) and guar (cluster bean) crop at Jodhpur indicated that relative humidity recorded beneath the tree canopy during the active cropping season of guar was found to be 7% more than in the open. This will, in turn, help for better growth of the crops.

Biodiversity conservation

Over exploitation of natural resources is a major challenge for sustainable production and livelihood security. Deforestation is the major cause which affected the biodiversity of an ecosystem. Agroforestry with components like trees, agricultural crops, grasses, livestock, etc. provides all kinds of life support system. However, agroforestry may not entirely reduce the deforestation but in many cases it acts as an effective buffer to deforestation. Trees in agroecosystems in Rajasthan and Uttarakhand have been found to support threatened cavity nesting birds and offer forage and habitat to many species of birds. These systems also act as a refuse to biodiversity after catastrophic events such as fire. The traditional society of coastal belts and tropics of the country practising homegardens and sacred groves help in bio-diversity conservation. Nonetheless, the traditional human dealings like beliefs, faith, taboos, customs and preferences must have played a central role in conservation of habitats and individual species (Jam, 20(X)). In a majority of the traditional villages, trees have been planted and dedicated to different gods/ goddesses or have stated as abodes of spirit, making them sacred. Frequently, species selected by the local people for social significance might turn out to be of ecological significance. It helps ensure sustainability, stability, and productivity of production systems in spite of the level of complexity of the ecosystem in which it occurs, and in the final analysis, it contributes to social welfare of the population through its involvement to poverty mitigation and sustainable food security. Similarly, the traditional people in hilly terrains of north-eastern India have been practising agroforestry in a variety of forms. The traditional homegardens are best example to it. As the people of north-eastern India have lived closely with the natural resources, their utility requirements also depend most on them. Hence, through trial and errors over generations, they have developed a kind of indigenous practice that involves domestication of tree species to meet the household requirements along with the vegetable crops in the home yards. Also, they have certain intuitive techniques of rotating the crops and/or recycling of kitchen wastes, etc. to maintain the soil fertility to a certain extent. Such a practice through the replenishment of soil fertility and the domestication of indigenous trees producing profitable forest products can be one way of diversifying and escalating the agro-ecosystem in a way that is positive to the environment and in maintaining and conceivably enhancing the biological diversity.

Carbon sequestration

Tree components in agroforestry systems can be significant sink of atmospheric carbon (C) due to their fast growth and high productivity. By including trees in agricultural production systems, agroforestry can, arguably, increase the amount of C stored in lands devoted to agriculture, while still allowing for the growing of food crops. In agroforestry systems, tree component is managed, often intensively by pruning for minimizing competition and maximize complementarity. The pruned materials are mostly non-timber products. Such materials often are returned to the soil. Besides, the amount of biomass and therefore C that is harvested and exported from the system is relatively low in relation to the productivity of the tree. Therefore, unlike in tree plantations and other monoculture systems, agroforestry seems to have unique advantage in terms of C sequestration.

The global contribution of agroforestry as a sink of C has been estimated in different eco-regions based on the tree growth rates and wood production, and assuming that the ratio of C in biomass is 50%, average C storage by agroforestry practices to be 9, 21, 50 and 63 Mg C/ha in semi-arid, sub-humid, humid and temperate regions. At global scale, agroforestry could be implemented on 585 to 1275 × 106ha and these lands could store 12 to 228 Mg C/ha tinder the prevalent climatic and edaphic conditions (Table 13.4).

In India, evidence is now emerging that agroforestry systems are promising land use system to increase and conserve above ground and soil C stocks to mitigate climate changes. The average potential of agroforestry has been estimated to be 25 t/C/ha over 96 m ha, 1998). In this way the total potential of agrotorestry in India to store C is about 2400m t. In another estimate (Fig. 13.7), the area under agroforestry is 8.2% of total reported geographical area (305.6 m ha) and it contribute 19.3% of total C stock under different land uses

(2755.5m t C). Although there is variation in the estimation of area under agroforestry and C stock made by scientist involve in this area but there is good indication of agroforestry for gaining popularity for mitigating climate change because desired tree cover can only be achieved including tree in farm field/bunds. The C storage capacity varied from region to region and also depend upon the growth and nature of tree species involved in the system (Table 13.5).

A study on C sequestration potential of *Albizia procera* under agri-silviculture was carried out during 2005 in a well established tree having 5-year-age with three pruning regimes (70% canopy pruning, 50% canopy pruning and un-pruned) with 2 crop rotations (blackgram-mustard and greengram-wheat) at NRC for Agroforestry, Jhansi. After 3-year, the system sequestered 23.58 to 24.79 t C/ha under different crop rotations irrespective of pruning regimes and the amount of C sequestered under pruning regimes was 27.97, 22.96 and 21.33 t C/ha in unpruned tree. 70% canopy pruning and 50% canopy pruning, respectively (Fig. 13.8). The C sequestration in pure tree and pure crop was 40 and 84% less, respectively than agrisilviculture. It indicates that agroforestry system have more potential to sequester C than either growing pure tree or pure crop. Tree without pruning can sequester more C but same time it does not allow cropping after 2- 3 years of planting. Canopy pruning of tree gives advantages of C sequestration as well as allow cropping for desired period.

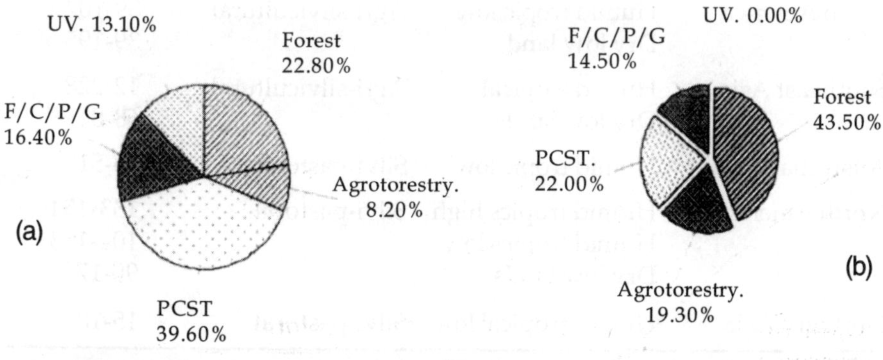

Note: PCST, Pure cropping with scattered trees in fields/on bunds, F/C/P/G, Fallow/cultivable wasteland/pasture/groves and UV, unfit for vegetation.

Fig. 13.7. (a) Per cent area under different land uses (total reported area 305.6 m ha). (b) Contribution of different land uses in carbon stock (total carbon stock 2755.5 mt).

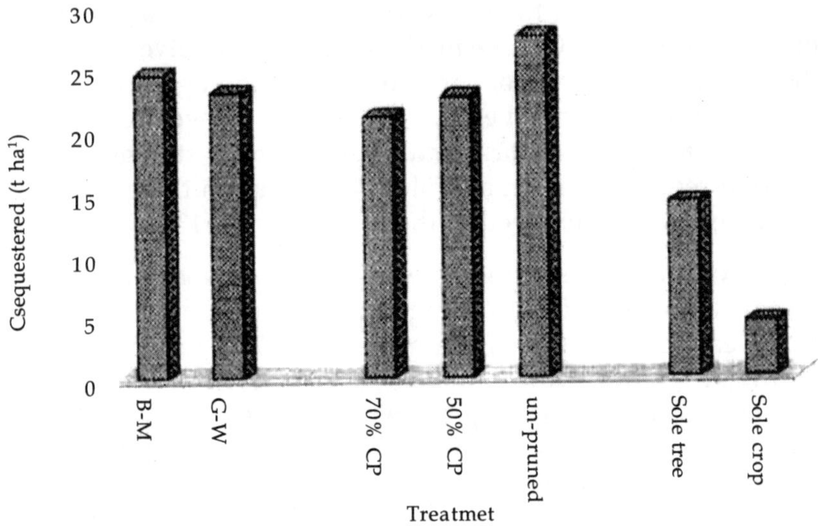

Note: B-M, Blackgram-mustard, G-W, Greengram-wheat, CP, Canopy pruning

Fig. 13.8. C sequestration potential of Albizia procera under different pruning regimes and crop rotation, sole tree and sole crop.

Table 13.4. Potential C storage for Agroforestry systems in different eco-regions of the world.

Country	Eco-region	System	Mg C/ha
Africa	Humid tropic high	Agri-silvicultural	29-53
South America	Humid tropic low Dry low land	Agri-silvicultural	39-102[a] 39-195
Southeast Asia	Humid tropical Dry low lands	Agri-silvicultural	12-228 68-81
Australia	Humid tropic low	Silvi-pastoral	28-51
North America	Humid tropics high Humid tropics low Dry low lands	Silvi-pastoral	133-154 104-198 90-175
Northern Asia	Humid tropical low	Silvi-pastoral	15-18

[a] C storage value were standardize to 50 year rotation.

Source: Dixon, 1995.

Table 13.5. Total C storage under agroforestry systems in different regions of the country.

Region	Agroforestry system and components	Total C storage (t C/ha)	References
Semi-arid region	Silvi-pastoral system (age 5 years)		
	Acacia nilotica + natural pasture	9.5-17.0	Rai *et al.*, (2001)
	A. nilotica + established pasture	19.7	
	Dalbergia sissoo + natural pasture	12.4	
	D. sissoo + established pasture	17.2	
	Hardwickia binata + natural pasture	16.2	
	H. binata + established pasture	17.0	
North-western India	Silvipastoral system (age 6 years)		
	Acacia/Dalbergia/Prosopis + *Desmostacya*	6.8-18.5	Kaur *et al.*, (2002)
	Acacia/Dalbergia/Prosopis + *Sporobolus*	1.5-12.3	
Central India	Block plantations (age 6 years)		
	Gmelina arborea	24.1-31.1	Swamy *et al.*, (2003)
Arid region (Rajasthan)	Agri-silvicultural system (age 8 years)		
	Emblica officinalis + *Vigna radiata*	12.7-13.0	Singh (2005)
	Hardwickia binata + *Vigna radiata*	8.6-8.8	
	Colophospermum mopane + *Vigna radiata*	4.7-5.3	
Semi-arid region	Agri-silvicultural system (age 11 yrs.)		
	Dalbergia sissoo + crop	26.0	NRCAF (2005)
North-western Himalayas	Silvi-pastoral system	2.17	
	Agri-horti-pastoral	1.15	
	Horti-pastoral	1.08	AICRPAF (2006)

Agroforestry for Bio-fuel and Bio-energy production

In rural areas 70-80% energy comes through biomass from trees and shrubs. Due to the agroforestry initiatives large amount of woods are now being produced from outside the conventional forest lands. Small landholdings and marginal farmers, through short rotation forestry and agroforestry practices are now providing the bulk of country's domestically produced timber products. Prosopisjuliflora is the major source of fuel for the boilers of the power generation plants in Andhra Pradesh (the other materials are rice husk, cotton stalks, other wood etc.). *P. juliflora* is widely distributed in Chittoor and Prakasam districts of Andhra Pradesh, Kutch-Bhuj district of Gujarat and Raichur and Bellary districts of Karnataka. The estimates made by the Forest Development Corporation, Gujarat on area of *P juliflora* in three districts of Gujarat *viz.*, Bhuj, Patan and Surendranagar was 0.17 m/ ha in 2005, with a potential of 0.15m t of charcoal production and employment generation of 1.855m mandays/rotation. The fuel wood potential of indigenous (*Acacia nilotica, Azadirachta indica, Casuarina equisetifolia, Dalbergia sissoo, Prosopis cineraria* and *Zyziphus mauntiana*) and exotic (*Acacia auriculiformis, A. tortilis, Eucalyptus camaldulensis* and *E. tereticornis*) trees revealed that calorific values ranges from 18.7 to 20.8 MJ/ kg for indigenous tree species and 17.3 tol9.3 MJ/kg for exotics. Species such as *C. equisetifolia, Prosopis juliflora, Leuceana leucocephala* and *Calliandra calothyrsus* have become prominent due to their potential for providing wood energy at the highest efficiency, shorter rotation and also their high adaptability to diverse habitats and climates.

In India the energy demand is expected to grow at 4.8%. Further in view of the increasing gap between demand and domestically produced petroleum the dependence on import of oil will increase in the near future. In this situation, fuels of biological origin have drawn a great deal of attention during the last two decades. Biofuels are renewable liquid fuels coming from biological raw materials and has proven to be good substitute for oil in the transportation sector as such biofuels are gaining worldwide acceptance as a solution for problems of environmental degradation, energy security, restricting imports, rural employment and agricultural economy. The potential tree borne oilseeds (TBOs) holding promise for biofuel are *Jatropha curcas, Pongamia pinnata, Simarouba, Azadirachta indica, Madhuca* spp., etc. In agroforestry system, *Jatropha curcas* has been intercropped with annual crops such as cowpea, sesame, sunflower, French bean, black gram, green gram and groundnut etc. at various places. The promotion of the use of oils could also provide a poverty alleviation option in the rural areas. Farmers can use vacant, waste and marginally used land for growing such trees and benefit from the annual produce, which will add as their income. With the increased green cover the environment will also benefit greatly. The use of oils is also CO_2 neutral, which would mitigate green house effect. But the economics and viability of the *Jatropha* plantation and bio-fuel production are still at initial stage and will be governed by international market prices of crude oil as well as government policies (NRCAF, 2007).

Challenges and opportunities

Due to increasing demands of food, fodder, fuel, timber and-for environmental security, the Indian agriculture is facing with lot of challenges and these are: inclusive growth and sustainable livelihoods, energy security, agricultural growth and food security and environmental security and climate change.

Agroforestry provides great opportunities to link water conservation with soil conservation; hence, the major focus has to be on this aspect. The opportunities offered by agroforestry are as mentioned here.

- Agroforestry systems create economic, environtnental and social capital, unleashing strong multipliers that contribute to making progress more inclusive and sustainable.
- Gives diversification.
- Provide opportunity for growth and also mechanism for better equity.
- Agroforestry provides societal continuum.
- Agroforestry is multifunctional system at the same time and thus, fulfills various demands simultaneously.
- Generate fresh water harvesting potential and ground water recharge.
- Agroforestry creates green cover for carbon sequestration.
- Regenerate biomass and nurtures depleted soil.

❑❑❑

Chapter-14

Biofertilizers and Sustainable Agriculture

In the last decades, synthetic fertilizers were used in modern agriculture. In the beginning farmers gets increased yield in agriculture. But slowly chemical fertilizers started displaying their ill-effects such *viz.* leaching out, polluting water bodies, destroying micro-organisms and friendly insects, making the crop more susceptible to the attack of diseases, reducing the soil fertility and thus causing irreparable damage to the overall system. Bio-fertilizer is a substance which contains live or latent cells of efficient strains of microbes which, when applied to seed, plant surfaces, or soil enhances the soil fertility by fixing atmospheric nitrogen, solublizing of phosphorus or decomposing organic wastes or augmenting plant growth by producing growth hormones with their biological activities.

In India, annual production of biofertilizer is increased up to around 40,000 tonnes (2011-12). The maximum biofertilizer is being produced in Tamil Nadu (11575.7tonnes) which contributes 32% of Indian biofertilizer production (NCOF, 2014)

Biofertilizers have important role in sustainable agriculture as stimulate plant growth, activate the soil biology, restore natural soil fertility, provide protection against drought and some soil borne diseases, increase crop yield by 20-30%, cost effective, supplements to fertilizers. They also play a vital role in sustainable agriculture as renewable source of nutrients, sustain soil health, supplement chemical fertilizers, replace 25-30% chemical fertilizers, increase the grain yields by 10-40%, decompose plant residues and stabilize C:N ratio of soil, improve structure and water holding capacity of soil. Biofertilizers as it replaces part of chemical fertilizer use reduces amount and cost of chemical fertilizers. They are an essential components of organic farming, play vital role in maintaining long term soil fertility and sustainability by fixing atmospheric nitrogen, combined application (N fixing and P solublizing) of biofertilizers increases the performance of both biofertilizers. Application of chemical fertilizers along with biofertilizers found more effective.

However, there is future need necessary to develop strains of bacteria suitable for different crops in different agroclimatic conditions and soil types to fully harness the benefit from the biofertilizers and there is need to put in greater effort in this direction to achieve better quality of inoculants by making necessary arrangements in storage and transport.

Biofertilizers play a key role in productivity and sustainability of soil and also protect the environment as it is ecofriendly and cost effective inputs for the farmers.

In the last century, chemical fertilizers were used in agriculture. Farmers were happy of getting increased yield in agriculture in the beginning.

But slowly chemical fertilizers started displaying their ill-effects such as :

- Leaching out
- Polluting water bodies
- Destroying micro-organisms and friendly insects
- Making the crop more susceptible to the attack of diseases
- Reducing the soil fertility and thus causing irreparable damage to the overall system.

Biofertilizers are groups of live and latent cell of microbes or micro-organism which are used to inoculate seed, plant surface and soil or all three which increase the production or productivity without impairing soil fertility.

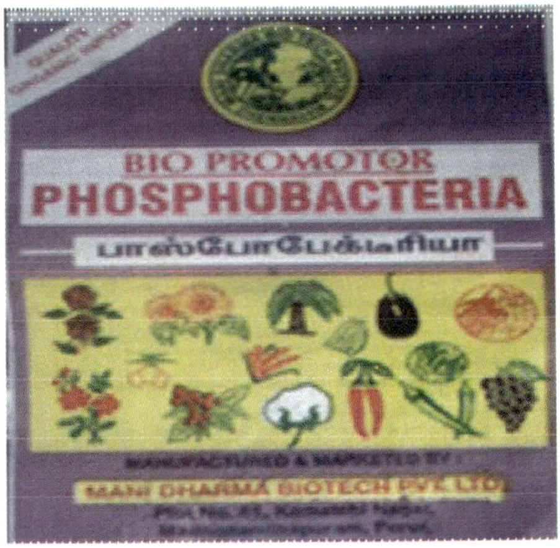

Concept of Biofertilizer

The term bio fertilizer refers to preparation containing live or latent cells of efficient strains of microbes which helps in enhancing the soil fertility by :

- Fixing atmospheric nitrogen,
- Solublization of phosphorus or
- Decomposing organic wastes or
- Augmenting plant growth by producing growth hormones with their biological activities.

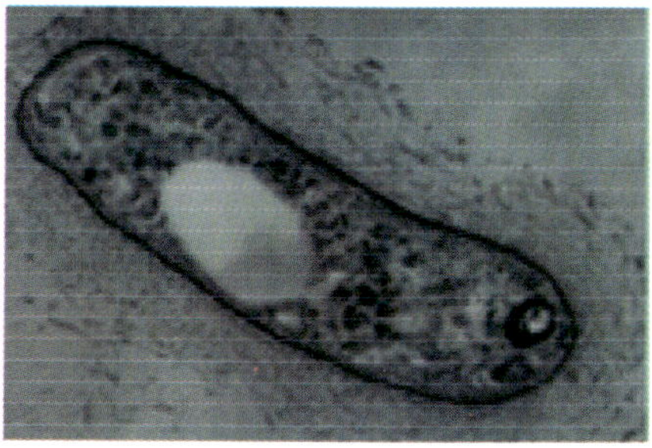

Rhizobium

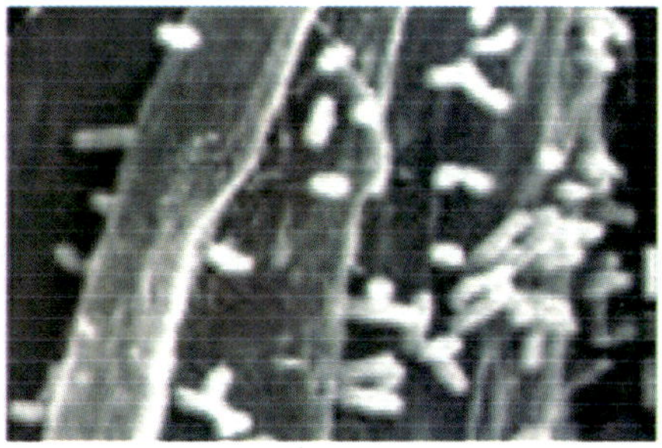

Bacteria in root

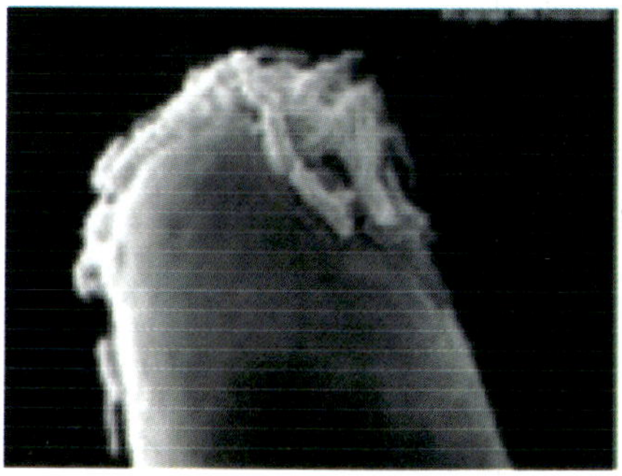

Bacteria in root

Legume

Benefits from using biofertilizers

- Stimulate plant growth.
- Activate the soil biologically.
- Restore natural soil fertility.
- Provide protection against drought and some soil borne diseases.
- Increase crop yield by 20-30%. Cost effective.
- Supplements to fertilizers.
- Ecofriendly (Friendly with nature).
- Reduces the costs towards fertilizers use, especially regarding nitrogen and phosphorus.

Role of biofertilizers for sustainable agriculture

- Renewable source of nutrients.
- Sustain soil health.
- Supplement chemical fertilizers.
- Replace 25-30% chemical fertilizers.

- Increase the grain yields upto 10-40%.
- Decompose plant residues and stabilize C:N ratio of soil.
- Improve structure and water holding capacity of soil.
- No adverse effect on plant growth and soil fertility.
- Stimulates plant growth by secreting growth hormones.
- Secrete fungistatic and antibiotic like substances.
- Solubilize and mobilize nutrients.
- Ecofriendly, non-pollutants and cost effective.
- Biofertilizers play a vital role in maintaining long term soil fertility and sustainability.
- Biofertilizers Production Scenario in India (in tonnes).

Biofertilizes Production Scenario in India (in tonnes)

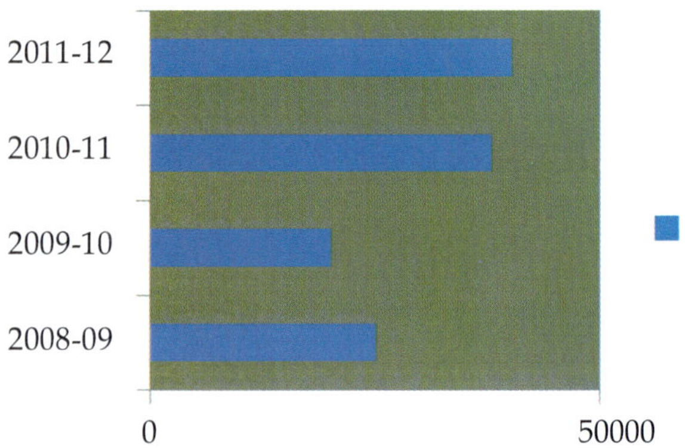

Production of biofertilizers in different states of India in tonnes (2012-13

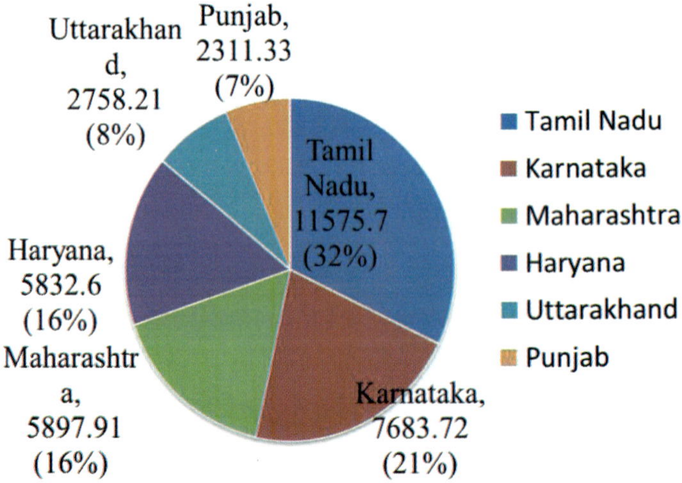

Source : National Centre of Organic Farming, Department of Agriculture and Cooperation, 2014

Classification of Biofertilizer

Biofertilizer Organisms

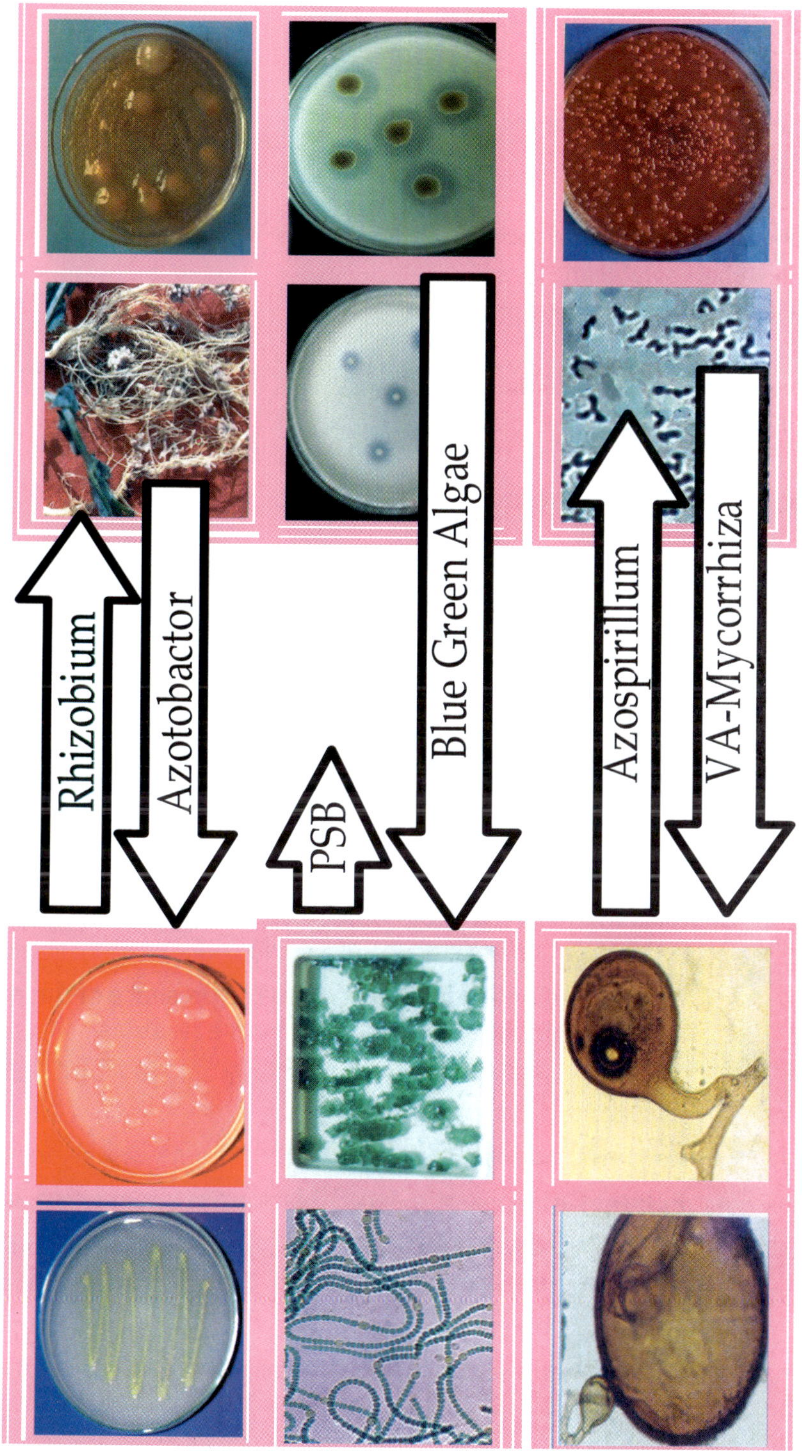

Suitable pH for microorganism

Microorganism	Optimum pH
Aerobic and heterotrophic	
Azotobacter	6.5 – 8.0
Beijerenkia	5.0 – 9.0
Phosphate solublizing bacteria	6.5 – 7.7
Phosphate solublizing fungi	5.5 – 7.0
Anaerobic and heterotrophic	
Clostridium	5.0 - 9.0
Nodule forming symbiotic	
Rhizobium	6.5 - 7.5
Frenkia	6.5 – 8.5
Non nodule forming symbiotic	
Azollae	5.5 - 7.0

Rhizobium

- That is 'slow growing rhizobia' known as ***Bradyrhizobium*** and the other group is 'fast growing rhizobia' called ***Rhizobium.***
- It is aerobic bacteria fixes atmospheric nitrogen in legumes symbiotically.
- ***Rhizobium*** lives in the root hairs of the legumes by forming nodules.
- *Rhizobium* enters the roots of legumes either through root hairs or at the point of emergence of lateral roots.
  - The pink coloured nodules due to the presence of the pigment leghaemoglobin are called as effective nodules.
  - This leghaemoglobin limits oxygen supply and helps the nitrogenase enzyme activity.
  - **Recommended for**
  - **Pulses** : Chickpea, pea, lentil, blackgram, greengram, cowpea, pigeonpea.
  - **Oilseeds**: soybean, groundnut.
  - **Fodders** : berseem, lucern.
  - Fixes 50-100 kg atmospheric nitrogen per hectare.
  - **Increase in yield : 10-35%**

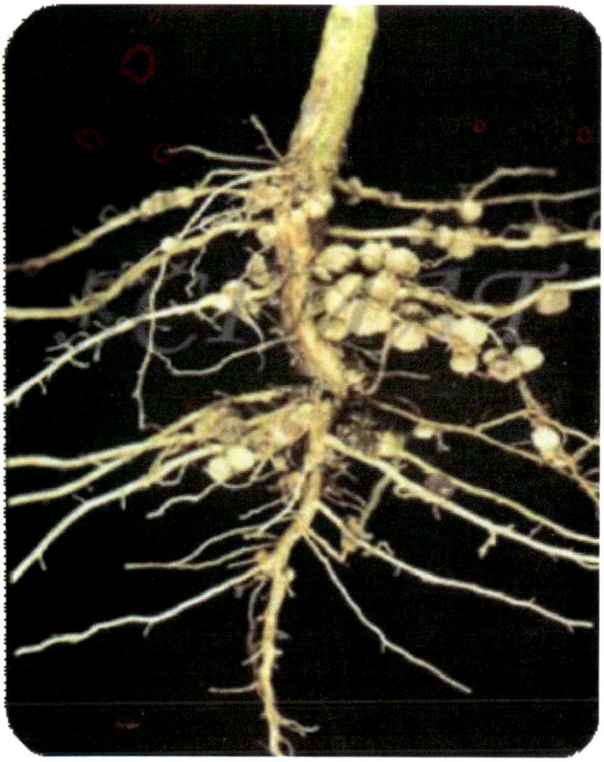

Genera and species of root nodule bacteria in legume

Genera **and** *Species*	**Legume crops**
Bradyrhizobium species	
B. japonicum	Soybean.
B. lupini	Lupine
Mesorhizobium species	
M. ciceri	Chickpea,
M. loti	*Trifolium*
Rhizobium species	
R. species	Moong, Redgram, Bengal gram, cowpea
R. leguminosarum	*Pisum* (pea), lentil.
R. phaseoli	Phaseolus (beans)
Sinorhizobium species	
S. meliloti	Alfalfa
S. saheli	*Sesbania*

Azolla

- *Azolla* is a free-floating water fern that floats in water.
- Fixes atmospheric nitrogen in association with blue green algae *Anabaena azollae.*
- *Azolla* is used as biofertilizer for wetland rice
- It contribute 40-60 kg N/ha per rice crop.

Azolla Application

The *Azolla* biofertilizer may be applied in two ways for the wetland paddy.

A. *Azolla* biomass incorporation as green manure for rice crop **(Befor transplanting)**

- Apply fresh *Azolla* biomass (15 tha^{-1}) to the main field and incorporate by using implements or tractor.

B. *Azolla* inoculation as dual crop for rice **(After Transplanting)**

- Broadcast the fresh *Azolla* in the transplanted rice field on 7th day after planting (500 kg / ha).
- Maintain water level at 5-7.5cm.
- A second bloom of *Azolla* will develop 8 weeks after transplanting which may be incorporated again.

Advantages of *Azolla* application

1. Accommodate P and other nutrients and releases slowly.
2. Can be applied as a dual crop in paddy fields.
3. Suppresses weeds.

4. Use as mulch, compost, fish feed, cattle feed, etc.
5. Edible as vegetable- *Azolla thoran*, *Azolla vada*, etc.- rich in protein.
6. Helps in detoxify heavy metals and many other poll.

Blue green algae (Cyanobacteria)

- BGA also called cyanobacteria
- They having **heterocyst** are capable of fixing atmospheric nitrogen.
- The most important species are *Anabaena* and *Nostoc*.
- These micro-organisms suitable under flooded rice.
- BGA applied at @ 10 kg per ha.
- About 25-30 kg N/ha could added in rice.

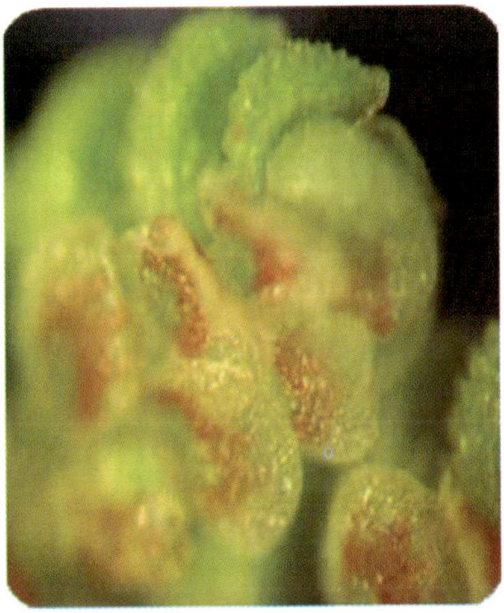

Blue green algae (Cyanobacteria)

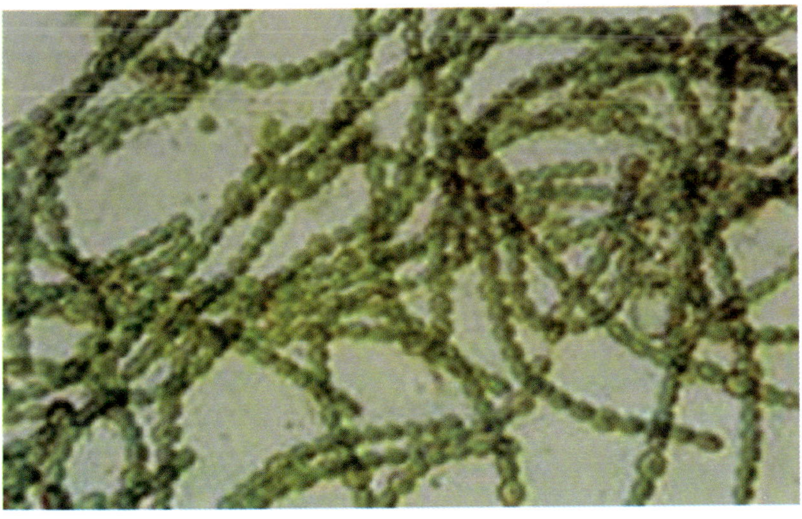

Phosphate Solubilizing Bacteria and Fungi

- A group of heterotrophic microorganism possesses the ability to solubilize organic P from **insoluble form to soluble forms.**

- **Bacteria:** *Bacillus megatherium, B. polymyxa, subtilis, Pseudomonas striata* etc.
- **Fungi:** *Aspergillus awamori, A. niger, Penicillum digitatum,* etc.
- **Recommended for :** All crops
- **Increase in yield: 10 to 20%**
- **Dosage**
- **Use 5 to 10 g/kg of seed, 1 to 2 kg for soil application, 1 kg for root application(root dipping) of one acre of crop.**

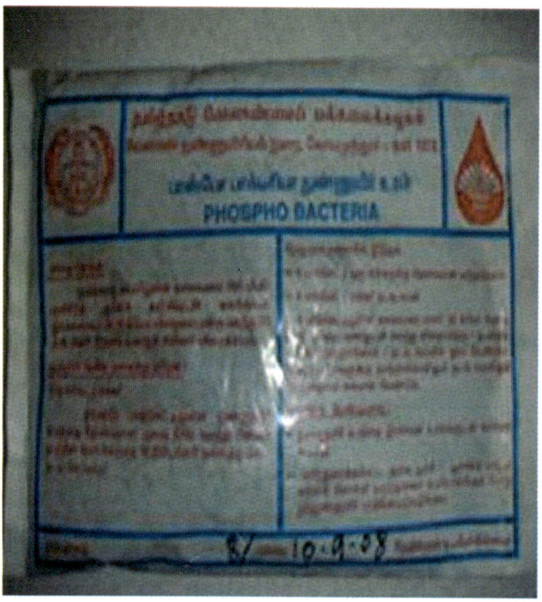

Vesicular Arbuscular Mycorrhiza (VAM)

- The term mycorrhiza was taken from Greek language meaning 'fungus root'.
- VAM, a fungus colonize the plant root system.
- As indicated above, the mycorrhiza is a mutualistic association between fungal mycelia and plant roots.
- VAM help in nutrient transfer mainly of phosphorus, zinc and sulphur.
- The major genus is *Glomus. spp.*

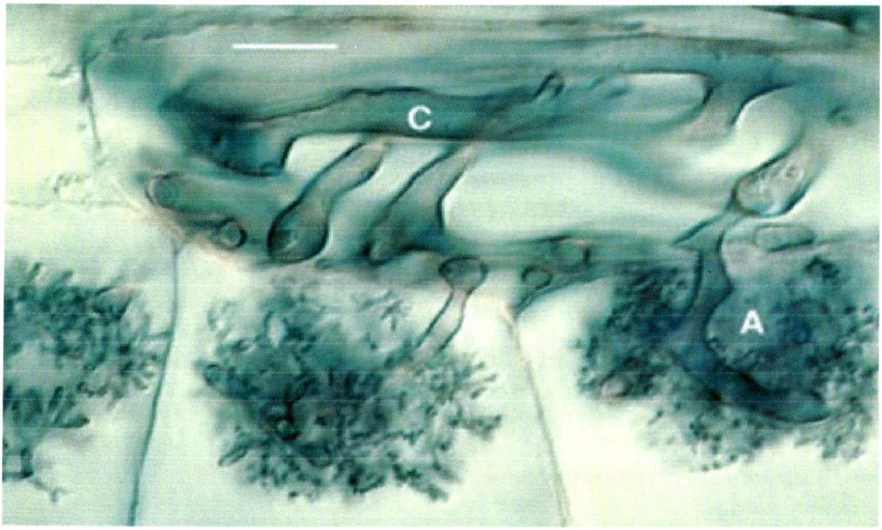

Actions of Mycorrhiza

- Enhances the feeding areas of the plant root is as the hyphae spreads around the roots.
- Mobilizes the nutrients from distance to root.
- Stores the nutrients (sp. phosphorus).
- Removes the toxic chemicals (example : phenolics)
- Provide protection against other fungi and nematodes.

Plant Growth Promoting Rhizobacteria

- The group of bacteria that colonize roots or rhizosphere soil and beneficial to crops are referred to as plant growth promoting Rzobacteria (PGPR).
- The PGPR inoculants currently commercialized that seem to promote growth through at least one mechanism,
- Suppression of plant disease (termed Bio protectants).
- Improved nutrient acquisition (termed Biofertilizers),
- or phytohormone production (termed Biostimulants).
- These PGPR are referred to as Bio stimulants and the phyto-hormones they produce indole-acetic acid, cytokinins, gibberellins and inhibitors of ethylene production.

Liquid Biofertilizers

- Liquid bio-fertilizers are special liquid formulation containing not only the desired microorganisms and their nutrients but also special cell protectants that promote formation of resting spores for longer shelf life and tolerance to adverse conditions. (Hegde, 2008).
- Bio-fertilizers manufactured in India are mostly carrier based and in the carrier-based (solid) bio-fertilizers, the microorganisms have a shelf life of only six months. They are not tolerant to UV rays and temperature more than 30 ^{0}C.

Advantages of Liquid Biofertilizer over conventional carrier based Bio-fertilizers

- Longer shelf life 12-24 months.
- No contamination.
- No loss of properties due to storage up to 45^{0} C.
- Easy identification by typical fermented smell.
- Better survival on seeds and soil.
- Very much easy to use by the farmer.
- Dosages is 10 time less than carrier based Bio-fertilizers.
- High commercial revenues.
- High export potential.
- Very high enzymatic activity.

Dosage and application of liquid Biofertilizers in different crops

Crop	Recommended Bio-fertilizer	Application method	Quantity to be used
Pulses, Chickpea, Pea, Groundnut, Soybean, Beans, Lentil, Lucern, Berseem, Green gram, Blackgram, Cowpea and Pigeonpea	*Rhizobium*	Seed treatment	200-500 ml/acre
Wheat, Pearlmillet, other cereals, Mastard, Sesamum Cotton, Castor and Vegetables.	*Azotobacter/ Azospirillum*	Seed treatment	200-500 ml/acre
Leguminous plants/ trees	*Rhizobium*	Soil treatment	1-2 ml/plant

Biofertilizer application by drip irrigation

Drip irrigation application, mix about one litre of liquid biofertilizer (Nitrofix) in 100 litres of chemicals free, good quality water. Filter the mixture with a pure cloth, use the filtered solution in drip irrigation for one acre.

Biofertilizers application by different methods

Seed treatment

Soil treatment

Root dipping

Drip Irrigation Application

Table 14.2. Dry pod weight, and kernel yield of groundnut as influenced by plant growth promoting *Rhizobacteria* under rainfed conditions.

Treatments	Dry pod weight (g plant^{-1})	Dry pod yield (kg ha^{-1})	Kernel yield (kg ha^{-1})
T_1 = Control	23.371	2597	1898
T_2 = $PGPR_1$	24.283	2797	2103
T_3 = $PGPR_2$	24.762	2834	2131
T_4 = $PGPR_4$	25.533	2858	2179
T_5 = $PGPR_1$ + $PGPR_2$ + $PGPR_4$	24.972	2914	2195
SEm±	0.460	34.99	27.86
LSD (0.05)	1.321	100	80

Table 14.3: Yield attributes and yield of rice as influenced by different N sources and their integration with *Azolla* and vermicompost.

Treatments	Effective tillers/m^2	Filled grains/ panicle	Grain yield (t/ha)	Straw yield (t/ha)
T_1 = Control	49.98	67.39	4.16	4.59
T_2 = V_{60}	52.71	73.65	4.63	5.00
T_3 = U_{40} + V_{20}	57.28	78.46	5.19	5.58
T_4 = U_{20} + V_{40}	55.15	75.55	4.91	5.45
T_5 = Azolla	53.23	73.90	4.61	5.00
T_6 = V_{60} + Azolla	53.35	75.19	4.74	5.18
T_7 = U_{20} + V_{40} + Azolla	55.86	76.65	4.98	5.50
T8 = U_{40} + V_{20} + Azolla	59.35	80.40	5.45	6.01
T_9 = U_{60} + Azolla	**60.50**	**83.98**	**5.70**	**6.24**
CD at 5 %	2.95	4.35	0.38	0.38

U–Urea; V- Vermicompost

Table 14.4: Response of *Azospirillum* inoculation on grain pot^{-1} (five hills pot^{-1}) in rice cultivars at two nitrogen levels. Each value is mean of three replicates in pots.

Cultivars	0 Kg N ha^{-1}			50 Kg N ha^{-1}		
	T_1	T_2	T_3	T_1	T_2	T_3
BPT=5204	65.70 (43.36)	72.00 (47.52)	73.30 (48.37)	80.00 (52.8)	85.30 (56.29)	87.70 (57.88)
Sarjoo-52	57.70 (36.10)	67.30 (44.41)	68.00 (44.88)	73.30 (48.37)	79.00 (52.14)	81.00 (53.46)
NDR-97	60.00 (39.60)	65.70 (43.36)	67.30 (44.41)	70.30 (46.39)	75.70 (49.96)	77.70 (51.28)
Pusa Basmati	35.00 (23.10)	39.30 (25.93)	41.00 (27.06)	42.30 (27.91)	48.70 (32.14)	50.00 (33.00)
MTU-7029	65.70 (43.36)	76.00 (50.16)	76.30 (50.35)	80.70 (53.26)	86.30 (56.15)	88.00 (58.06)
HUR-36	50.70 (33.46)	57.00 (37.62)	57.3 (37.81)	59.30 (39.13)	65.70 (43.36)	67.00 (44.22)

CD at P 0.05 = 2.45 g. Value in parenthesis denotes value of grain yield q ha^{-1} (pot size = 0.15 sq.m.)

T_1 = Control (without *Azospirillum* coating); T_2 = Seed coating with standard *Azospirillum* Biofertilizer strain (IARI); T_3 = Seed coating with local isolate strain of *Azospirillum*

Table 14.5. Effect of seed treatment with biofertilizer on yield, weight and quality of *Arachis hypogea*

Treatment	Pods/plant	Pod yield/ plant (g)	Protein content (%)	Oil content(%)
T_1= Control(NPK)	19.5	44.0	22.1	51.12
T_2= A+B+ *Azotobacter*	20.0	37.1	22.7	52.08
T_3 = A+B+ BGA	22.4	48.8	23.4	53.28
T_4= A+B+ *Azospirillum*	21.3	53.0	23.2	53.48
T_5= A+B+ *Azospirillum* + *Azotobacter*	21.8	54.5	23.0	52.1
T_6= A+B+ + BGA+ *Azotobacter*	24.0	60.9	24.2	52.64
T_7= A+B+ + BGA+ *Azospirillum*	27.4	90.5	25.8	56.46
T_8=A+B+BGA+ *Azotobacter+Azospirillum*	**30.5**	**108.7**	**26.3**	**57.2**

Tamil Nadu, A+B= 75% FYM + 25% NPK. Sujanya *et al.* 2011

Table 14.6: Biofertilizer inoculation effect on seed cotton yield and yield parameters of cotton.

Treatments	Seed cotton Yield (kg ha^{-1})		No. of bolls plant^{-1}	
kg P ha^{-1}	Un-inoculated	inoculated	Un- inoculated	inoculated
T_1= 30	1377.7	1489.0	29	34
T_2= 60	1544.3	1666.7	32	36
T_3= 90	1611.3	1733.3	32	37
LSD (0.05)	107.25		3.489	

Faisalabad

Table 14.7: Growth, Yield and Oil % of sunflower as influenced by biofertilizers. (Pooled data 2 year)

Treatments	Seed yields (kg ha-^{1})	Biological yield (kg ha-^{1})	No. of filled seed capitulum	Oil yield (kg ha-^{1})	Test weight (gm)
T_1 = *Azotobacter*	1969	7512	418.5	622	47.11
T_2 = PSB + *Azotobacter*	2342	8587	471.1	780	49.05
T_3 = VAM + *Azotobacter*	2648	9216	507.3	918	50.13
T_4 = PSB + VAM + *Azotobacter*	3225	10289	567.6	1210	53.84
SEm (±)	**44**	**103**	**4.6**	**14**	**0.40**
CD at 5 %	**127**	**297**	**13.2**	**40**	**1.15**

West Bengal

Table 14.8: Effect of bio-fertilizer inoculation on physiochemical parameters of soil (*Vigna mungo)*

Treatment	pH	Available N (kg/ha)	vailable P (kg/ha)	Available K (kg/ha)
Before sowing	7.58	113.5	4.7	99.7
After harvesting				
T_1 = Control	7.60	116.2	5.8	110.3
T_2 = *Rhizobium*	7.34	158.2	8.1	121.2
T_3 = *Phosphobacteria*	7.41	151.1	9.7	123.1
T_4 = *Rhizobium* + *Phosphobacteria*	7.33	159.7	9.02	124.7

Jhansi, U.P.

Table 14.9: Effect of INM on grain and straw yield of rice and wheat

Treatments	Rice yield (t/ha.)		Wheat yield (t/ha.)	
	Grain	Straw	Grain	Straw
T_1 =Control	2.05	2.93	1.74	2.00
T_2 = Farmer practice (N100, P 40, K0)	3.68	5.33	3.00	3,78
T_3=T_3 100 % NPK Zn.	4.73	6.58	3.30	4.28
T_4 = T_3 + FYM 5 t/ha	5.11	7.16	3.48	4.44
T_5= T_3 + PSB	5.01	7.13	3.37	4.32
T_6= T_3 + BGA/*Azotobacter*	5.07	7.24	3.41	4.36
T_7=T_3 + BGA/*Azotobacter* + PSB	5.16	7.35	3.46	4.40
T_8= T_3 + BGA/*Azotobacter* + PSB + FYM 5 t/ha	5.46	7.87	3.57	4.62
SEm±	0.12	0.18	0.15	0.09
CD at 5 %	0.24	0.36	0.30	0.19

Kanpur . U. P., Bahadur *et al.* (2013)

Table 14.10: Effect of bio-fertilizers on yield and quality of groundnut

Treatments	Kernel yield (kg/ha)	Pod yield (kg/ha)	Haulm yield (kg/ha)	Biological Yield (kg/ha)	Oil content (%)	Protein Content (%)
T_1 = Control	1897	2923	4787	7709	47.80	24.91
T_2= *Rhizobium*	2182	3170	5060	8230	48.80	25.56
T_3 = PSB	2084	3069	4934	8003	49.07	25.37
CD at 5%	92	80	157	181	0.95	0.31

Jaipur (Rajasthan), Pareek *et al.* (2013)

Table 14.11: Effect of bio-fertilizer application on growth, yield parameters and quality parameters of greengram

Treatments	Number of pods plant-	Number of seeds pod-1	Seed yield (kg ha-1)	Stalk yield (kg ha-1)	Protein content (%)
Biofertilizer Uninoculated	16.34	9.65	1274.68	2000.72	20.12
PSB (Bio.) Inoculated	17.39	10.35	1350.19	2153.29	20.57
S.Em.±	0.336	0.187	26.359	44.318	0.124
CD at 5 %	0.954	0.531	74.917	125.959	0.351

Sardarkrushinagar, Gujarat, Patel *et al.* (2013)

Table 14.12: Quality and yield parameters of soybean as influenced by levels and sources of P and biofertilizers.

Treatments	Protein content (%)	Oil yield (t/ha)	Seed yield (t/ha)	Harvest Index (%)
T_1= Control	36.29	0.20	1.10	29.42
T_2=PSB + VAM	37.03	0.22	1.21	29.33
T_3=50% P as RP	37.14	0.22	1.23	30.17
T_4=50% P as RP + PSB + VAM	38.76	0.27	1.48	32.47
T_5=50% P as DAP	37.73	0.25	1.38	31.04
T_6=50% P as DAP + PSB + VAM	37.45	**0.28**	1.51	32.00
T_7=100% P as RP	37.72	0.26	1.40	30.46
T_8=100% P as RP + PSB + VAM	**38.89**	**0.28**	1.56	35.33
T9=100% P as DAP + PSB + VAM	38.86	0.28	1.55	34.35
SEm±	0.18	0.004	0.02	0.41
CD (P=0.05)	0.52	0.01	0.06	1.18

IARI, New Delhi

Table 14.13. The effect of organic manures & biofertilizer combinations on grain, straw yield and protein content of wheat.

Treatment	Grain yield (t ha^{-1})	Straw yield (t ha^{-1})	Protein content (%)
T_1 = Control	2.4	4.9	6.99
T_2 = Farmyard manure (FYM)	3.4	5.7	8.47
T_3 = Green manure (GM)	3.8	6.0	8.65
T_4 = GM + biofertilizers (B)	4.1	6.3	8.74
T_5 = GM + FYM + B	4.6	6.6	9.26
SEd±	0.19	0.17	0.42
CD at 5 %	0.39	0.34	0.86

Pali, Rajasthan

Probable reasons for getting poor responses of biofertilizers

- Ineffective microbial strain
- Insufficient microbial population
- High levels of contaminants
- May have been exposed to high temperatures or sun light

- Not following recommended methods and dosage
- Using simultaneously with inorganic fertilizers or pesticides
- Prevailing high soil temperatures and low soil moisture
- Acidity or Alkalinity in soil

For improving effectiveness of bioferilizers

Biofertilizers should be stored in a cool place or at room temperature (25°C to 28°C for longer shelf life). The simplest method is to store in a pitcher containing wet paddy straw.

- Avoid direct contact with chemical fertilizers, insecticides and pesticides during storage or application
- The storage and use of biofertilizers should be avoided in direct sunlight.
- The inoculant should be used before the expiry date.
- When fungicides and insecticides are to be used, apply fungicides before insecticides application. Apply biofertilizers after all other treatments, give a gap of 24 hrs.
- *Rhizobium* biofertilizer is specific to host crop, hence, only specific inoculant should be used for specific crop.
- Biofertilizers as it replaces part of chemical fertilizer use, reduces amount and cost of chemical fertilizers.
- Bio-fertilizers being essential components of organic farming play vital role in maintaining long term soil fertility and sustainability by fixing atmospheric nitrogen
- Combined application (N fixing and P solublizing) of biofertilizers increases the performance of both biofertilizers.
- Application of chemical fertilizers along with biofertilizers found more effective.
- Biofertilizers would play key role in productivity and sustainability of soil and also protect the environment as it is eco-friendly and cost effective inputs for the farmers.

Future Needs

- There is necessary to develop strains of bacteria suitable for different crops in different agro climatic conditions and soil types to fully harness the benefit from the biofertilizers.
- There is need to put in greater effort in this direction to achieve better quality of inoculants by making necessary arrangements in storage and transport.
- The effect of biofertilizers on crop growth and yield is not as striking as that of chemical fertilizers. However, on long term sustainability basis, biofertilizers has an important role.

❑❑❑

Chapter-15

Mushroom Production Technology

Button Mushroom (*Agaricus bisporus*)

The white button mushroom [*Agaricus bisporus* (Lang) Sing.] is very popular throughout the world and is the most important on commercial scale. This mushroom is extensively cultivated throughout the world and contributed about 60 per cent of the total world production of the mushroom. At present, India is roughly producing about 50,000 tonnes of this mushroom. It can be successfully cultivated in place where the environmental conditions are favourable. The optimum temperature for mycelial growth is 20-25° C and that for fruit body fonnation 15-20° C and also needs a high percentage of relative humidity (85-95%). Cultivation of white button mushroom requires technical skill.

Formulation of Compost

A. Compost Formulation used in India

a) B.H.U. Varanasi

1.	Wheat straw	600 kg
2.	Wheat bran (chokar)	60 kg
3.	Urea	7.5 kg
4.	Murate of potash	6.0 kg
5.	Single super phosphate	6.0 kg
6.	Molasses(rab)	9.0 kg
7.	Gypsum	60 kg
8.	Insecticide Powder	1/2 kg

b) DNIR-Solan

1. Wheat straw	1000kg
2. Chicken manure	400 kg
3. Brewers grain	72kg
4. Urea	14.5kg
5. Gypsum	30kg
6. B.H.C. dust powder (10%)	1/2 kg

c) I.A.R.I. - New Delhi

1. Horse yard manure	1000 kg
2. Wheat straw	500 kg
3. Urea	30-40 kg
4. Gypsum	30kg

d) I.I.H.R. - Bengaluru

1. Wheat straw or paddy straw	500 kg
2. Wheat bran (chokar)	30 kg
3. Urea	12 kg
4. Calcium ammonium nitrate	6.0 kg
5. Single super phosphate	9.0 kg
6. Calcium carbonate	10 kg
7. Gypsum	12 kg

e) Punjab Agriculture University- - Ludhiana

1. Wheat straw +paddy straw (1: 1)	300 kg
2. Wheat bran (chokar)	3 kg
3. Urea	3 kg
4. Murate of potash	3.0 kg
5. Single super phosphate	6.0 kg
6. Molasses (rab)	5 kg
7. Gypsum	3 kg
8. Calcium ammonium nitrate	9 kg
9. B.H.C. granular (10%)	250 gm

f) Rajendra Agriculture University- Pusa (Samastipur)

1.	Wheat straw + paddy straw (1:1)	500 kg
2.	Chicken manure	150 kg
3.	Urea	10 kg
4.	Murate of potash	5 kg
5.	Single super phosphate	10 kg
6.	Calcium carbonate	1 kg
7.	Gypsum	20 kg
8.	Calcium ammonium nitrate	9 kg
9.	Calcium Sulphate	1 kg
10.	Bavistin	50 gm
11.	Formalin	250 ml
12.	DDVP	1 liter

g) Sonepat Area

1.	Wheat straw	1000 kg
2.	Wheat bran (chokar)	30 kg
3.	Urea	7.5kg
4.	Murate of potash	3.0 kg
5.	Single super phosphate	7.5 kg
6.	Molasses (rab)	23 kg
7.	Gypsum	30 kg
8.	Calcium ammonium nitrate	7.5 kg
9.	B.H.C. granular (10%)	770 gm
10.	Furadon	770 gm

h) South India

1.	HorseDung	-100 kg
2.	Urea	6.8 kg

3.	Rice Bran	20 kg
4.	Gypsum	25 kg

OR

1.	Paddy straw	1000 kg
2.	Urea	10 kg
3.	Rice Bran	100 kg
4.	Cotton seed	12kg
5.	Gypsum	24 kg

B. Compost Formulation used abroad

a) Netherlands

1.	Horse Dung	100 kg
2.	Chicken manure	100 kg
3.	Urea	15 kg
4.	Gypsum	25 kg

b) Korea

1.	Rice straw	1000 kg
2.	Chicken manure	100 kg
3.	Urea	10-12 kg
4.	Gypsum	10-20 kg

c) Taiwan

1.	Rice straw	1000 kg
2.	Ammonium sulphate	20 kg
3.	Super phosphate	20 kg
4.	Urea	5 kg
5.	Chalk	30 kg

Preparation of compost: The compost is prepared by long method of composting (LMC) which was first advocated by Mantel *et al*. 1972. If pasteurization facilities are available, then the short method of composting (SMC) is applied, however, where pasteurization facilities are lacking, the long method of composting is applied.

A. Short method of composting

A few progressive growers are using pasteurized compost prepared by short method ofcomposting. It is done in two phases taking 10 days and 7 days respectively.

Phase-I, Outdoor composting.

Phase-II, Indoor composting in peak rooms or bulk pasteurization tunnels.

Phase-I, Outdoor composting : This phase start with the wetting of raw materials. Wheat straw and chicken manure are mixed thoroughly and wetted properly till they absorb sufficient water. Leached water is regularly resprayed over the ingredients. After thorough wetting, a heap is made out of such materials by trampling it hard to encourage anaerobic fermentation. After 2 days and 4 days the stack is broken by first and second turning respectively. Gypsum is added in compost materials during third turning after 6 days. After 8 days the compost is filled in the pasteurization/peak heating room or bulk chambers. The phase first culminates on the 8th day when the compost is filled for pasteurization. In phase-1, the ingredients are allowed to decompose outdoor under uncontrol conditions, where manure is partially decomposed. Temperature buildup spontaneously in the stack in the inner most compost.Temperature of the zone outer to inner most layer range between 50 and 60°C. This zone harbour huge population of thermophilic mieroflora which decompose the ingredients in this layer. The layer also serves as the inoculum for the thermophilic flora when compost is turned.

Phase-II, Indoor composting in peak healing rooms or bulk pasteurization tunnels : After above process compost is loosely filled manually or by filling machine in the tunnel up to the height of 5-6 feet. Filling is so that 900-1000 kg of compost occupies about one squre meter of space. Several up down temperature sensors are place inside the tunnel. One sensor is below the compost for air temperature. All the door are closed after the filling and blower fan is switched on to bring the compost to uniform temperature at 45° C. The difference in compost temperature and air temperature above should not be more than 3°C. When the compost temperature reach 45° C, fresh air is stopped and compost temperature start rising by self-generation of head and it may rise at the rate 1.2° C/hour. Required temperature needed for the pasteurization of compost at 57° C may reach in 10-12 hours. Some quantity of steam can be used for this purpose. When pasteurization is complete, compost is ready for conditioning.

Conditioning: The compost is brought to a temperature that is optimum for the majority of thermophilic flora. These organisms convert ammonia in to microbial protein. Excess of ammonia is released into atmosphere. The thermophilic microorganisms which develop in to compost, are utilized by the mushroom mycelium for its growth at 25°C when they become almost inactive. Compost is conditioned at 40°C because some mesophilics develop at this temperature which may render compost unsuitable for mushroom growth. Further if such mesophilic flora persist in the compost at spawning, it may lead to rise in compost temperature, killing the mushroom mycelium. Beside proper temperature supply of fresh oxygen is very important during conditioning as it is an aerobic process.

B. Long method of composting

Wheat straw is spread on cemented floor and wetted thoroughly by sprinkling water. This mosit straw mixed with other ingredients and again wetted with water. A heap (stack) is made of this mixture and covered with polythene sheet. The compost materials are fermented by total seven turnings and each turning is done at 4 days interval. The heap is made after each turning and it should not be compressed tightly otherwise anaerobic condition may be created in the compost heap. Gypsum is mixed during 3rd turning and BHC dust powder mixed at the last turning for prevention of insect pests. At each turning water should be sprinkled to make up the loss of moisture content due to evaporation. The compost is ready for spawning, if it is dark brown in colour, without any smell of ammonia and has sufficient moisture rontent (68-70%) when pressed between palm.

Spawning

Spawning is the mixing of spawn in the well prepared compost. For spawning, completely colonized and fresh spawn should be used. The amount of spawn is 2-2.5 kg./quintal compost. The spawned : ompost @ 5 kg. is filled in one polythene bag. The upper surface of : ompost is covered with news paper sheet within the polythene bag. These spawned bags are placed in growing chamber where temperature anged between 20-25°C.

Casing

After spawn run, the mycelium become ready to enter the eproductive phase to produce the fruit bodies (mushroom). However, t will not produce fruit bodies unless the compost surface is covered, 'he process of applying casing layer on the surface of the composted is called "casing of compost beds". Casing also provides moisture or growth and development of fruit bodies and maintain humidity and temperature in the growing chamber.

After completion of spawn run, the newspaper sheet is removed nd surface of compost is covered (3-4cm thick layer) by casing soil, he casing soil is prepared from 2 years old farm yard manure and loam soil (1:1 ratio). The casing materials are wetted with water and terilized by autoclaving or formalin solution before casing. Mushroom beds are sprayed regularly with water to keep the casing layer dequately moist.

Harvesting

Pinheads generally appears within 15 days after casing and they ecome ready for harvesting within another one week. The mature nopened mushroom buttons are harvested at most once a day. They re picked by gently twisting of the button without easing disturbance, mall pits are formed after harvesting and these pits are immediately re-eased with easing soil for development moisture in the easing soil is maintained by regular spraying of water.

Yield

The yield of white button mushroom depends on the compost ingredients, supplements used in the compost, easing materials, temperature and relative humidity. 15-20kg fresh button mushroom can be raised from one quintal compost.

Precautions

1. Wheat straw should be dry and free from rotting and inert matters.
2. Completely colonized and fresh spawn should be used for spawning.
3. Spraying water should not be excessive and stay on casing surface.

Button Mushroom

C. Oyster Mushroom (*Pleurotus* species)

Oyster mushroom (*Pleurolus* spp.) have become a well recognized part of our country and world mushroom industries. The desirable attributes like fast mycelial growth, high ability for saprophytic colonization, very simple and cheap cultivation technique, higher yield and easy post harvest storage and above all the wide choice of *Pleurotus* sp. available for cultivation under different climatic condition have contributed to its popularity.

Huge quantities of lingo-cellulosic crop residues and other organic wastes are generated annually through the activities of agricultural, forest and food processing industries. According to the estimates of Food and Agricultural Organization (FAO, 1989) about 2946 million tonnes of cereals straws are produced worldwide of which Asia's share is 1135 million tonnes and that of India is 240 million tonnes. In India, during 1993-94 production of paddy straw and wheat straw was approximated at 110.6 million tonnes and 153.6 million tonnes respectively (Mathew 1996). Recently about 385 million tonnes of agricultural wastes are available annually in India and about half of this residue remains unused. The major parts of these residues are burnt after havesting, resulting multifaceted hazards including oxygen deficient enviornment, respiratory diseases (such as allergic, bronchitis, asthma, tuberculosis) and poor visibility at night. If even one per cent of this crop residue is used to produce mushrooms, India will become a major mushroom producing country in the world.

Requirements

- Mushroom spawn
- Agricultural residues (cereal straw, wooden parts etc.)
- Polythene bags
- Sterilization equipment/chemicals
- Plastic drum
- Wooden tray

Preparation of Substrates

Oyster mushroom (*Pleurotus* sp.) can be grown on various agricultural waste materials. They grow well on cereal straw, sugarcane, bagasse, cotton waste, jute coir pith, water hyacinth, mango stiff, sunflower straw, pulse straw, groundnut pod shell, some weed straw, rubber wood dust, grasses etc. However, wheat straw and paddy straw are best substrates for cultivation of oyster mushroom and they give higher yield.

These agricultural residues should be chopped and dried and free from disease, rotting, contamination and other inert matters. For preparation of substrates, cereal straw is soaked in water for 18-20 hours and then drain off the excess water. Various methods of substrates sterilization have been developed for *Pleurtous* cultivation. Steam pasteurization by autoclave, hot water treatment, chemical sterilization and solarization are effective methods and they are adopted by mushroom growers on largt* scale.

Autoclaving: The moist substrates are steamed under pressure inside the autoclave for 20 minutes at 10 lb. After pasteurization substrates are taken out of the autoclave and allowed to cool down at room temperature.

Chemical sterilization: The pre-soaked substrates are again immersed in 0.05% water solution of Bavistin for 12 hours and then excess solution is drain off from the substrates. Another chemcial sterilization is moistened substrates fumigated with formalin for 12 hours in air-tight chamber/plastic drum. These substrates are exposed to air after fumigation for one hour to get escape from the odour of formalin.

Solarization: Solarization technique is more effective for cultivation for milky mushroom because temperature is higher in this cultivation period. This is also effective for cultivation of oyster mushroom but sun-light should be properly raising during process.

In solarization, the moistened straw are spread on cemented floor in thin layer and covered with clean polythene sheet and treated in sun-light from 10 a.m. to 4 p.m.

Spawning: The mixing of spawn (mushroom seed) in the sterilized substrate is called spawning. The amount of spawn is 3-3.5 kg per quintal substrate. The spawning is usually done by mixing and layer methods. The cylindrical polythene bags (56x45cm) can be used for spawning. The spawned substrate is filled in polythene bags and mouth of each bag is bind with rubber band/threads and 8-10 small holes (2mm) are made at 10 cm apart rom each other.

Cropping chamber The spawned bags are kept vertically in a dark room where temperature ranged between 24-28°C. In Oyster mushroom, complete spawn running usually take place in 15-20 days, if above temperature and relative humidity (80-85%) are favourable.

Oyster Mushroom

After completion of spawn run, polythene bags are tear off and remove the side without disturbing the beds. This compact block is kept on the wooden racks.

Water spraying: To provide adequate moisture, daily water spray on the mushroom beds is essential. Water spraying should not be excessive and retain on the substrate.

Harvesting: Usually 3-4 days after opening the bags, mushroom primordia (fruit bodies) begain to form. Fruit bodies can be harvested in about 3 days after their appearance. Harvesting is done by grasping the stalk and gently pulling or twisting the mushroom from the substrate level. Mushroom fruiting is continued after harvesting of 1st flush. The bags should be removed from the mushroom house when substrate becomes colourfess and soft.

Yield: The yield of oyster mushroom is depending on (he kinds of utilized substrates, combination of substrate, supplements used in the substrate, temperature and relative humidity during cropping. Yield ranged from 50-100 kg per quinta dry weight of substrate.

Precautions

1. For preparation of substrate, cereal straw should be chopped, dry and free from rotting and inert matters.
2. Very old and contaminated spawn never used for spawning.
3. Spraying water should not be excessive and stay on mushroom beds.

□□□

Chapter-16

Multipurpose Trees and Shrubs

In agroforestry systems, it is necessary and economical if selected trees and shrubs are having their various uses so that each and every part, produce and product (s) of the trees are fully utilized for domestic purpose, industrial purpose etc. Besides, multipurpose trees and shrubs help in protecting enironment. They also become an integral part of other land uses (i.e. for amenity-shade, shelter, ornamental and windbreaks; lmprovment in soil through leaf fall, nitrogen fixation etc.

As mentioned above, multipurpose trees and shrubs (one or the other) provide the following items for domestic and industrial uses.

Domestic and Industrial Uses

- Fuelwood for cooking foods and warming houses,
- Timber, posts and poles,
- Pulp for paper cardboards, and construction boards,
- Vegetables, fruits and nuts, seeds, firewood.
- Edible leaves and shoots for sauces, curries, salads and beverages,
- Honey for medicinal uses,
- Fodder for livestock,
- Host plant for silkworms and lac insects,
- Green manure for incorporating into soil,
- Tan bark for the tannins used in tendering and dyeing leather in leather industry,

- Medicines and pharmaceuticals,
- Extractives such as a gums, resins, rubber and dyes,
- Oil (edible), and non-edible for illumination, lubrication and soap making,
- Cake obtained after oil extraction for the use of animal feed, if edible and non-edible types are for the use as manure,
- The fruit of aonla, harad and bahera are used as soap substitutes,
- Essential oil, perfumes are for cosmetics.

Environmental Protection and other uses

- Shade and support for plantation crops.
- Aid to the soil improvement by cotrolling soil erosion and sand dunes stabilization.
- Improvement in soil structure, chemical properties, water balance and weed control.
- Useful in hedges, windbreaks, shefterbelts, and fire breaks.
- Ornamental, aesthetic or recreatidnal value.

A few multipurpose tree species for arid and semi-arid regions

Botanical Name	Common Name	Economic Importance
Acaicia albida	African Savanna tree	Fodder, firewood, tannin, gum
Acacia aneura	Mulga acacia	Fodder, firewood
Acacia nilotica Delile	Babul, kikar	Fodder, tannin, gum, small, timber, farm, implements, tool handles, construction, work, fence post cross-arms, mallet heads, anvil blocks, tent pins, pegs, ladder rungs
Acacia senegal Willd	Gum Arabic acacia	Firewood, gum, soil conservation
Acacia tortilis Forsk	Israeli babul	Fodder, firewood, gum, farm implements tender immature seeds are used as vegetable
Azadirachta indica A. Juss	Neem	Timber construction farm implements seed oil, cake for manure, tannin, firewood, gum, medicinal, bee flora

Cajanus cajan indica A. Juss	Pigeonpea, Arhar	Firewood, food, forage, amenity planting
Capparis deciduas Pav	Kair	Fruits, timber firewood, medicinal
Cassia siamea Lam	Kassod, kassid	Flower, edible gum green manure, pulp, for paper, medicinal, tannin, timber, mallet heads, walking stick, ornamental
Casucarina equisetifolia L	Saru	Timber, pole, firewood, ballies, fence posts, cross-arms, charcoal, tannin, pulp for paper, resin, medicinal
Dalbergia sissoo Roxb	Shisham	Timber, farm implements, charcoal, fodder, wood oil for skin disease treatment
Emblica officinal is gaertn	Aonla	Fruit, tannin, firewood, fodder, oil, medicinal, furniture, farm implements
Eucalyptus cama-ldulensis Dehnh.	River Redgum	Paper pulp, box making, oil, bee flora, firewood, poles and posts for building construction, anvil blocks
E. gomphocephala A.D.C.	Tuart	-do-
E. microtheca F. Muell		-do-
E. teretocornis Sm.	Forest Redgum	-do-
Gleditsia triacanthes	Honey locust	Fodder, firewood, soil conservation, farm implements
Gmelina arborea Roxb	Gambhar	Small timber, box making, firewood, charcoal, paper pulp, fodder, fruit edible, medicinal
Leucaena leucocep-hala (Lamk) de Wit	Subabul	Fodder, biomars, farm implements, small timber fence, posts, pegs, tool handles, paper pulp, charcoal
Madhuca indica J.F. Gmel	Mahua	Fruit, fodder, oil (fatty), paper pulp, timber,mallet heads, doors and window, agricultural implements, peg, tent pins handles, medicinal
Pithecellobium dulce Benth	Manila Tamarind	Fruit, fodder, firewood, seed oil, seed meal, small timber, farm - implements, tannin, medicinal

Prosopis cineraria L	Khejri	Fodder, firewood, small- timber, farm implements, medicinal, bee flora, paper pulp
Prosopis julifora DC	Mesquite	Fodder, firewood, small- timber, farm implements, tool handles, tent pins, peg, fence posts, tannin, bee flora, paper pulp, gum, medicinal
Salvadora oleoides Decne Pilu		Fruits, seed fat or soap and candle making, deoiled cake for animal feed, manure, timber, fuelwood, medicinal
Tamarix aphylla Karst.	Tamarisk	Small timber, farm- implements, firewood basket making, tannin
Zizyphus mauritiana	Lam Bordi	Fruit, fodder, firewood
Zizyphus nummularia	Wight & Arn	Farm implements, charcoal, dye and tannin from bark

A few multipurpose tree species for humid tropical regions

Botanical Name	Common Name	Economic Importance
Acacia auriculiformis A. Cunn	Australian wattle	Small limber, firewood, charcoal, Tannin, ornamental
Albizia lebbeck Benth	Siris	Fodder, firewood, charcoal, tannin, gum, timber, farm implements, manure, fodder
Anacardium occidentale L	Cashew	Fruit, kernels, oil, timber, firewood
Anthocephalus chinensis A. rich	Kadam	Fodder, fruit, pulp- paper, firewood
Areca catechu L.	Supari	Nut, chewing gum, tannin, fat, firewood, manure, mulching, pulp paper medicinal
Artocarpus hetero-phyllus Lamk.	Jackfruit	Fruit, vegetable, Fodder, small limber, packing boxes, plywood, malletheads, furniture, ladders, farm implements, firewood, medicinal
Azadirachta indica A. Juss	Neem	As mentioned ealier

Bambusa spp.	Bans	Pulp paper, basket making, pickle from tender shoots, small timber, poles.
Cassia siamea Lam	Kassod	As mentioned earlier
Casuarinas equisetifolia L.	Saru	As mentioned earlier
Cocos nucifera L.	Coconut	Coconut water, copra, coconut oil, coconut-cala, toddy, jaggery, coir, wood for small construction, leaves for thatch, jhaps, leaf petioles and rachis for broco ins
Cordial dichotoma Forst	Indian cherry	Fruits,small timber, fire-wood, fodder, tannin medicinal
Eucalyptus spp.	Eucalyptus	As mentioned earlier
Gliricidia sepium	Giliricidia	Green manaure, mulching-material, firewood, ornamental
Grevilea robusta A. Cunn.	Silver oak	Timber, green manure, firewood.
Leucaena leucocephala	Subabul	As mentioned earlier
Mangifera indica L.	Mango	Fruit, timber, firewood, fodder, seedfat medicinal
Morus alba L.	Mulberry	Fodder, silkworm rearing, firewood, fruits, small- timber, basket making
Pithecellobium dulce	Manila Tamarind	As has mentioned
Pongamia pinnaa Pierre	Karanj	Oil for burning, soap making and medicinal uses, oil cake manure, leaf manure, Timber, for construction, tool- handles, agricultural implements, firewood, fruits
Sesbania grandiflora Ruiz Pav.	Agast	Flower, fruits & leaves used as vegetable, firewood fodder, green manure, tannin, gum
Syzygium cumini Skeels	Jaman	Fruits, products are syrub, jelly, jam, wine, vinegar, pickles, tannin, paper,tassar silk wornus rearing,bee flora
Terminalia catappa L.	Indian almond	Timber, firewood fruit,gum lannin, medicinal

A few multipurpose tree species for temperate regions

Botanical Name	Common Name	Economic Importance
Acacia dealbata Link	Silver wattle	Gum, flower are used in perfumery, tannin and dye, ornamental
Acacia decurrens Willd	Green wattle	Timber, firewood, soil conservation, ornamental
Acer oblongum Wall	Maple	Fodder, firewood
Albizia lebbeck Benth	Siris	As mentioned earlier
Alnus nitida Endl	Arder	Timber, tannin, ornamental
Aesculus indica Colebr.	Indian Horse	Timber and fruits chestnut
Bombax ceiba	Semal	Timber, firewood, fodder, feed, silky flosss, gum, medicinal
Carya illionensis	Pican nut	Fruit and timber
Castanea sativa Mill	Chestnut	Fruit, fodder and timber
Celtis australis L.	Nettle	Timber, firewood fodder
Eucalyptus citriodora		Essential oil (cineole)
Eucalyptus globulus Labill	Blue gum	Timber, firewood, bee flora paper, medicinal
Grevillea robusta A. cunn	Silveroak	Ornamental, gum, green-manure, Honey
Grewi aoptiva Drummond	Bhimal	Small timber, firewood, fodder, fruits, fibre, bee flora
Juglans regia L.	Walnut	Fruits, fodder, timber, firewood, dye.
Melia azadirachta L.	Bakain	Fodder, timber, firewood, seed oil, medicinal, bee flora
Populus spp.	Poplar	Timber, paper pulp, firewood, fodder, match industry
Prunus cornuta Steud	Bird cherry	Fruit and fodder Jamunol
Robinia pseudacacia L.	False Acacia, Robinia	Small timber, farm implements firewood, fodder, bee flora, seed can be used as cattle feed after boiling.
Sapindus mukorossi Gaertn	Soapnut, Aritha	Detergent, Kernel oil in soap industry, wood as fuel and making charcoal.
Terminalia arjuna Bedd	Arjuna	As mentioned earlier

❑❑❑

Chapter-17

Nursery Management

For better Nursery Management, the following essentials should be taken care of:

- Meaning of nursery
- Nursery bed
- Nursery stock
- Importance of nursery
- Types of nursery
- Site for nursery
- Area of nursery
- Fencing of nursery area
- Layout of nursery
- Dimensions of nursery bed
- Soil and its preparation
- Treatment of nursery soil
- Seed arrangement
- Pre-treatment of seed
- Sowing of seeds in nursery
    - i) Methods of sowing
    - ii) Seed rate

iii) Time of sowing

iv) Distance and depth of sowing

- Mulching of nursery bed
- Irrigation of nursery bed
- Protection of seedlings in nursery bed.
- Interculture and weeding
- Shifting
- Pricking
- Thinning

Meaning of nursery

The nursery means an area where plants are raised and after certain period of time they are shifted to other planting place.

Nursery-bed

The nursery bed is a well prepared area in a nursery where seeds are sown or transplants or cutting are planted for certain period of time. There are mainly three types of beds i.e. flat bed, raised bed and sunken bed. Flat bed is easy to prepare but water logging cannot be prevented during rains. Raised bed is preferred during rains. Sunken bed is used for placing seedling along with containers during summers and scarcity of water.

Nursery stock

All planting material which is grown in nursery for planting purpose is knwon as nursery stock.

Importance of nursery

- The importance of nursery lies in the preparation of healthy and vigorous plant stock.
- The plant species which usually do not produce seed can be made available for planting every year. The seeds of such plant species can be collected during the seed year and raised in the nursery. The seedlings from the nursery can be planted in poor seed year.

The seedhngs in the nursery bed are maintained by shifting them in the second nursery and to third nursery.

- The plant species which are very slow growing when they are sown directly in the permanent places are most likely to be suppressed by weeds and killed. Therefore, slow growing species are generally raised in nursery and planted out only when the seedlings have attained plantable size.
- Success of road side avenue plantations, depends largely on planting tall and sturdy plants. Tall seedlings can only be obtained when they are raised in the nursery for one year or two.
- The plant species having very small seeds are liable to be blown or washed, therefore, raising them in the nursery becomes essential.
- Many times, germination of seed is low (30-40%) and difficult in poor, barren and wastelands while direct sowing yields variable results. Therefore, it becomes essential to raise seedlings in the nursery. On difficult sites, mortality of seedlings is very common. Gap filling by seedlings is better than by seeds because nursery grown plants grow quick and establish early. They are also not harmed much by weeds. There will be no much age difference between gap filled plants and the main plants (which were planted earlier).Therefore, it is essential to grow plants in nursery for gap filling.
- It is necessary to raise rare and expensive planting material in the nursery because effective supervision is possible in nursery for watering, weeding and control for pest and diseases.

Types of nursery

There are mainly two basis for classification of nurseries. These basis are irrigation facilities and duration of the use. Based on irrigation facilities, nurseries are of two types such as dry nursery and wet nursery.

Dry nursery

It is nursery that is maintained without irrigation.

Wet nursery

It is a nursery that is maintained with irrigation.

Based on duration of the use, nurseries are of two types such as temporary nursery and permanent nursery.

Temporary nursery

It is for short period and intended to meet the requirement of a limited area.

Merits

- It dose not need much manuring because temporary nursery is made on rich cleared sites.
- It enables easy, quick and cheap transporting of planting stock without any serious damage or shock.

Demerits

In the absence of irrigation facility, the growth of seedlings is usually slow and there is usually heavy mortality in them. It is relatively costlier to maintain as compared to permanent nursery.

Permanent nursery

It is a nursery that is meant for supply of nursery stock for a long time. Its area is large. It, generally has irrigation and shading facilities for plants.

Merit

Its site remains near central place and hence supervision is easy and effective. It should be on or near some road for quick transport of nursery stock.

Site for nursery

- As far as possible, nursery area should be located centrally so that supply of seedlings becomes easy for plantation.
- It should be nearer to water supply, labours etc.
- For better growth of seedlings, soil should be light and well drained.
- Nursery bed should in open, but there should be some side shade from the south-west side.
- The nursery area should be free from the serious insect pests and diseases.

Area of nursery

The area of nursery may vary from 0.5 to 2.5 per cent of the plantation area depending upon:

- Plant species
- Climatic conditions
- Soil types
- Irrigation facilities
- Age of planting of seedlings.

Fencing of nursery area

In order to protect nursery from animals, it is necessary to fence the nursery area. In dry regions live hedge is raised with a dense row of shrubs like *Acacia nilotica, Pithecellobium dulce, Duranta* etc. This live fence will also help in preventing soil erosion and the effect of hot winds.

Layout of nursery

First of all, whole area should be demarketed and then, it should be divided into rectangular or square blocks by permanent paths of about 1.5 to 1.8m width. An individual block is further divided by paths of 0.9m width into small blocks which are sub-divided into rectangular nursery beds by paths of 0.5 rn width. Along paths, the provision of irrigation channels is also made.

In the hills, first of all terracing is done and then nursery beds are prepared. The lengthwise direction of the beds should be east to west that will help in protecting nursery from frost and sun through shading east and west direction.

Dimensions of nursery bed

The size of beds varies from region to region depending upon climatic conditions, kind and type of soil (topography etc.), area of plantation etc. However, the nursery beds of 10 × 1m are commonly adoptable size.

Soil and its Preparation

Soil should be dug up to a depth of 0.5 metre and all stone, pebbles, stumps, etc should be collected. This should be followed by a cross harrowing. Farmyard manure (2-3 kg/sqm) should be incorporated during land preparetion. About 75 kg Aldex 5 per cent dust should be mixed in the soil to protect seedlings from white ants.

During rainy season, it is advisable to raise the level of beds at least 15-20 cm for good drainage. In order to prevent sides of the beds from crumbling in dry season and from eroding during rainy season the sides of the beds are supported with bamboos in a line of bricks, stones etc.

Treatment of nursery soil

As far as possible, soil of nursery should be free from pests and disease pathogens. Generally, termites pose problems in many nursery soils. To prevent damage from termites, 5 per cent Aldex, Melathion etc. dust at the rate of 75 kg per hectare is applied. Pests can also be controlled by burning of dry leaves and organic debris on the surface of prepared nursery bed after sowing of seeds and transplanting of seedlings. Nursery soils are also treated with formalin to prevent spread of certain pathogens.

Seed arrangement

For successful nursery management, obtaining quality seed is most important. One should find out the reliable source of getting seed material. One should have the knowledge of how and when to collect seed, how to extract seed and method of storage.

Source of seeds

There are mainly three sources from which seed can be procured. They are : (i) Seed agencies like Government Departments, Agricultural Universities, Forest Departments/Institutions, State Seed Farms etc.(ii) Private Seed Growers, and (iii) Self Collection.

Seed orchards

The Seeds are usually collected from forests, seed orchards and scattered growing trees.

Seed orchards are of two types: (i) Clonal, and (ii) Seedling seed orchard. Clonal orchard is raised from the plants propagated from common parent by vegetative method(s).

Clone means all cuttings and grafts derived from plants obtained by sowing seeds from plus trees.

Advantages of clonal orchards

- Plants come in bearing earlier,
- They are genetically superior, and
- Their yield potential is higher.

Seed collection should be done as and when ripen fruits are available. During spring and summer, a large number of tree species produce seeds. Seed is also obtained during winter in a considerable number of species. The season during which generally seed is obtained in a particular species may be early or late due to local climatic conditions and soil types. The year to year variation in seed period may also be due to variation in climate.

How to collect seed

Seed can be collected by mainly three ways:

- Seed is collected from the ground which are either naturally fallen or they are made to tall down by shaking the branches of the tree. For example *Acacia spp., Lasora.*
- Seed is collected from lopped branches and/or felled tree. For example *Terminalia myriocarpa, Betula, Dalbergia sissoo, Acacia catechu* etc.
- Seed is collected from the fruit (pod) plucked from the standing tree. For example *Acer, Morus* etc.

Whereever applicable seeds are extracted from the fruits that follow drying, cleaning and grading. Both graded or ungraded seeds are kept in polythene bags or glass or metallic containers and stored in a dry and clean place.

Pre-Treatment of Seed

The seeds of many plant species need treatment due to their hard seed coat and dormancy due to other reasons. Treatment enhances better germination. Some kind of seeds are soaked in cold water or boiling water and some other kinds are treated with chemicals. (viz., lime water, dilute alkali, dilute acid, salt solutions etc.).

Hot water treatment of seeds

This method of seed treatment is very easy and convenient to follow. Not only this, a large number of hard seed coated tree species respond well to it. First of all, in a container tap water is measured to cover all the seeds to be treated to about double their depth. Thereafter, water is allowed to get heated to boiling point without seed. When boiling, water is removed from the heat, and it is allowed to cool for about 4-5 minutes.

It will be better it the temperature of the hot water is known through thermometer, it is cooled until 85°C. Then water is poured on the seeds to be treated and stirred thoroughly for about 3 or 4 minutes, so that all the seeds get exposed to hot water. No harm if seeds are allowed to soak for over night or so. After soaking, seeds are taken out from the water or water is allowed to drain off from the seeds, and seeds are spread on polythene sheet or gunny cloth under shade for one or two hours before sowing. Seeds can be dressed with fungicide(s) also wherever applicable.

The main objectives for treatment of seeds in hot water are:

i) Hot water breaks down the waxy, resistant layer of the seed coat,

ii) Embryo gets adequate moisture and it becomes activated for germination, and

iii) Some inhibitors are also leached out from the seeds.

Acid treatment of seeds

Similar to hot water treatmeht, acid treatment of hard coated seeds is also performed particularly with sulphuric acid (H_2SO_4). Seed treated with acid germinate very early and in higher percentage. For example, *Acacia nilotica* seeds when treated with sulphuric acid for 1 to 1.5 hours results in higher percentage of germination. Not only this, germination starts within 2 days of acid treatment.

Sowing of seed in nursery

Method of sowing

Always seeds should be sown on well prepared nursery bed; the surface soil should be fine. There should not be any stone, roots, stumps etc. The seed can be sown by broadcasting or drilling or by dibbling. Precaution should be taken with very small seeds that they should be sown after mixing with fine sand or woodash for uniform distribution and for regulating desired seed rate. The depth of sowing for all the kind of seeds should be optimum. Deeper sown seeds may not easily germinate and also germinating sprouts may not come up properly due to the formation of soil crust. On the contray, there would also not be much shallow sowing that the seeds are eaten by rats, birds and insects. The usual practice is to have a soil thickness above the seed equal to the diameter of the seeds. The depth of sowing and distance between seed to seed and row to row is given in below.

Time, distance and depth of seed sowing of a few tree species in the nursery bed

Species	Time of sowing	Distance of sowing in cm (row × seed)	Depth of sowing (cm)
Acacia catechu	May-June	20×10	1.5
Acacia niotica	April-June	20×10	1.5
Adina cordifolia	April-June	15×10	0.5
Ailanthus excelsa	March-May	20×10	0.5
Ailanthus pendula	March-May	15×15	0.5
Albizia chinensis	June-July	15×10	1.0
Albizia lebbeck	June-July	15×15	1.0
Anogeissus pendula	June-July	15×10	0.5
Azadirachta indica	June-August	20×10	1.0-1.5
Bombax ceiba	April-June	15×5	0.5
Butea monosperma	June-July	15×5	0.5
Casuarinas equisetifolia	March-April	15×5	0.5
Celtis australis	Feb.-April	20×5	1.5-2.0
Dalbergia sissoo	March-April	25×2	1.0-1.5
Eucalyptus tereticornis	May-June	10×5	0.5
Gmelina arborea	May-June	15×10	1.0-1.5
Grevillia robusta	Nov.-Dec.	20×10	1.0
Grewia optiva	March-April	25×5	1.5-2.0
Holoptelia integrifolia	June-July	10×10	0.5
Leucaena leucocephala	June-July	30×5	1.0-1.5
Madhuca longifolia	July-August	15×3	1.5-2.5
Manilkara hexandra	April-May	30×10	1.0
Melia azedarach	Feb.-March	15×10	2.0
Moringa oleifera	May-June	20×5	1.0-1.5

Morus alba	May-June	20×5	0.5
Pinus roxburghii	Sept.-Oct.	20×10	1.0
Pithecellobium dulce	April-May	30×10	2.0
Pongamia pinnata	June-July	15×2	2.0
Prosopis cineraria	June-July	15×10	1.5-2.0
Prosopis juliflora	June-July	15×10	1.5-2.0
Quercus glauca	Feb.-March	15×10	1.0
Robinia pseudacacia	March-July	20×5	1.0
Sesbania grandiflora	May-June	20×5	1.0-1.5
Shorea robusta	June-July	20×10	2.0-3.0
Syzygium cumini	May-July	30×10	2.0
Tamarindus indica	March-April	20×10	2.0-3.0
Terminalia arjuna	May-July	30×10	2.0-3.0
Terminalia bellerica	March-April	30×10	2.0
Terminalia catappa	July-August	30×10	2.0-3.0
Terminalia chebula	June-July	30×10	2.0-3.0
Toona ciliate	June-July	20×5	0.5
Zizyphus nummularia	May-June	30×10	1.0-1.5

Number of seeds per kilogramme of some common plantation species

Species	**Seeds per kilogramme**	
Acacia albida	11,500	20,000
Acacia auriculiformis A. cunn.	40,000	
Acacia catechu Willd.	38,000	40,000
Acacia dealbata L.	71,000	80,000
Acacia decurrens Willd	80,000	1,11,000,00
Acacia farnesiana Willd	9,000	16,000
Acacia mearnsii de Willd	65,000	75,000
Acacia melanoxylon R.Br.	55,000	94,000
Acacia mellifera Benth	20,000	
Acacia nilotica Delite	12,000	15,000
Acacia Senegal Willd	71,000	33,000
Acacia tortilis Hayne	12,000	16,000

Acrocarpus fraxinifolius	26,000	30,000
Adenanthera pavonina L.	28,000	51,000
Aegle marnelos Correa	53,000	
Ailanthus altissima Swingle	35,000	1,00,000
Ailanthus excelsa Roxb	95,000	
Albizia amara Boiv	13,500	14,500
Albizia falcateria	38,000	44,000
Albizia lebbeck Benth	8,000	11,000
Albizia procera Benth	20,000	22,000
Alnus napallnsis D. Don.	21,800	5,70,000
Anacardum occidentale L.	150	500
Annona squamosa L.	2,800	4,500
Anthocephalus chinensis	Several Million	
Araucaria cunnighamaii	2,400	4,000
Azadirachta indica	4,000	4,500
Barringtonia acutangula Gaertn	1,400	
Bauhinia racemosa Lam.	7,000	9,000
Bauhinia purpurea L.	4,000	5,000
Bauhinia semla Wenderlin	1,450	1,600
Bauhinia variegata L.	2,800	3,200
Betula alnoides Buch-Ham	93,00,000	20,000,000
Bischofia javanica Blume	58,400	104,000
Bombax ceiba L.	21,400	38,500
Butea monosperna (Lam) Taub	9,850	14,700
Calophyllum inophyllum L.	100	211
Cassia fistula L.	2,600	7,100
Cassia siamea	37,000	45,000
Casuarinas equisetifolia L.	7,00,000	8,00,000
Cedrela odorata	45,000	60,000
Cordial alliodora	20,000	30,000
Cordial dichotoma Forst	13,00	14,00
Cupressus lusilanica	1,70,000	3,20,000
Dalbergia latifolia Roxb.	18,500	40,000
Dalbergia sissoo Roxb	13,000	16,000
Dodonaea viscose (L) Jacq	70,000	1,00,000
Emblica officinalis Gaertn.	68,000	8,90,000

Eucalyptus camaldulensis Dehnh	7,00,000	8,00,000
Eucalyptus deglupta Blume	Several Million	
Eucalyptus grandis W. Hill	6,00,000	7,00,000
Eucalyptus maidenii F. Muell	3,00,000	3,50,000
Eucalyptus tereticornis Sm.	3,00,000	8,00,000
Gmelian arborea Roxb.	700	1,400
Grevillea robusta A. cunn.	1,00,000	1,05,000
Holoptelea integrifolia Planch.	9,000	32,000
Jatropha curcas L.	22,000	2,500
Juglans regia L.	90	-
Leucaena leucocephala (lamk) de Wit	27,000	35,000
Madhuca longifolia Macb.	350	450
Melia azedarach L.	1,400	1,800
Moringa oleifera lam	6,000	9,000
Oujeinia oojeinensis	28,000	
Parkinsonia aculeate L.	12,000	13,000
Phoenix dactylifera L.	750	-
Pinus caribaea More	52,000	72,000
Pinuselliottii Engelm	30,000	37,000
Pinus gerardiana wall	30,000	32,000
Pinus kesiya Royle	55,000	62,000
Pinus markusii Jungh & de Vriese	50,000	62,000
Pinus oocuarpa	41,000	55,000
Pinus patula Sch. Champ	1,00,000	1,60,000
Pithecellobium dulce Benth	6,700	10,000
Prosopis chilensis Stuntz	9,800	36,000
Prosopis cineraria Druece	25,000	27,000
Robinia pseudacacia L.	30,000	40,000
Salvadora oleoides Decne	13,000	-
Samanea saman Merrill	6,450	-
Sesbanea bispinosa W. F. Wight	1,000	16,000
Simmondsia chinensis Schneid	700	5,000
Syzygium cumini Skeels	1,200	-
Swietenia macrophylla	1,300	3,700
Tamarindus indica	1,000	1,800
Tamarix aphylla Karst	2,86,000	-

Tecomella undulata Seem	70,200	-
Tectona grandis L.	2,000	3,100
Terminalia alata Heyne	13,000	-
Terminalia arjuna Wight & AM.	750	800
Terminalia catappa L.	100	-
Thespesia populnea Soland	6,300	-
Treema orientalis Blume	3,60,000	-
Zizyphus mauritiana Lam	1,200	1,700

Seed rate

The quantity of seed depends upon many factors viz, kind of seed (small or big), climatic condition, soil types, viability of seeds, purity of seeds, method of sowing (broad-casting, drilling, dibbling etc).

Species of which seed is produced in different states in India

State	Name of species
Andra Pradesh	*Eucalyptus hybrid, Tectona grandis,*
Bihar	*Eucalyptus hybrid, Tectona grandis,*
Himachal Pradesh	*Pinus roxburghii*
Madhya Pradesh	*Tectona grandis*
Orissa	*Casuarinas equisetifolia, Eucalyptus hybrid,*
Uttar Pradesh	*Acacia catechu, Adina cordifolia, Bombax ceiba, Cedrus deodara, Dalbergia sissoo, Eucalyptus hybrid, Holoptelia integrifolia, Morus alba, Pinus roxburghii, Tectona grandis, Terminalia bellerica, Toonna ciliate*
West Bengal	*Abies densa, Acrocarpus fraxinifolius, Alnus nepalensis, Betula alnoides, Bombax Ceiba, Chukrasia velutina, Cryptomeria japonica, Cupressus cashmeriana, Duabanga grandifflora, Eucalyptus hybrid, Gmelina arborea, Lagerstroemia speciosa, Michelia champaca, Pinus patula, Pinus roxburghii, Schima wallichii, Shorea robusta, Tectona grandis, Terminalia myriocarpa, Tsuga dumosa*

The quantity of seed can be roughly calculated by the following formula:

$$W = \frac{A \times D}{P \times N} \times 100$$

Where,

W = Weight of seed required in grams.

A = Bed area in sqm.

D = Number of plants required per sqm.

P = Plant per cent of the species.

N = Number of seeds per gram.

Practically, twice the quantity of seed should be used for drill sowing and six time for broadcast sowing. The approximate number of seeds in one kg weight is given indicates species of which seed is produced in different states in India.

Time of Sowing

The time of sowing of seeds in the nursery bed depends upon the time of seed ripening, the duration of viability of seeds (short or long), the growth rate of species, and the size of the plant to be planted out etc. For instance, when the size of the plant to be planted out is small and the seed ripens in winter, in such case, sowing of the seed in spring season can give plants, fit for plantrg during rainy season. But when tall plants are required for planting out or stumps are prepared for planting, in such case, seeds are sown in the onset of rainy season and plants attain their suitable plantable size by the next rainy season. The duration or plants retaining in the nursery bed depends on their rate of the growth which can be regulated by manipulation of nursery management techniques like sowing date, application of water, (quantity and interval), addition of manure and fertilizers (quantity particularly), multiple transplanting, pruning of roots and/or shoots etc. However, it can be adjusted according to climatic conditions of the locality and transplanting time at the permanent sites.

Distance and depth of sowing

Generally, seeds are sown at narrow distance in the nursery so that there are less spots in the event of poor germination of seed. Closer spacing is also adopted as less area is required. Further, depth of sowing of seeds depends on type of soil and size of seeds.

Mulching of Nursery bed

Just after sowing of the seeds in the nursery sometimes, it becomes necessary to mulch it with the help of dry grasses. Mulching will conserve moisture and help inermination of seeds. Mulching material should be removed as soon as seeds start germinating and sprouts coming out from the ground.

Irrigation of Nursery bed

It will be better if nursery bed is given presowing irrigation so that sufficient moisture is available for germination of seeds. After sowing of the seeds, nursery beds should be supplied with very light and frequent irrigations preferably with watercan or sprinklers. The sprinkling system of irrigation is very efficient. This system has added advantages that when it is in operation, leaves of plants are washed of dust, dirt and certain salts and gases deposited on the surface of leaves due to air-pollution. Washed leaves may become more capable of performing active photosynthesis which may results in formation of adequate photosynthates for better growth and development of plants and ultimately higher yields.

Protection of Seedlings in Nursery bed

Young seedlings require protection from

a) Intense sun,

b) frost,

c) Heavy rainfall, hail storm

d) Pests and diseases, and

e) Browsing animals.

By erecting temporary shade with the help of wooden posts or sticks fixed in the ground covered with thatch, split bamboo, plant twigs which are locally available. Young seedlings can be protected from the damage causes due to intense sun, frost, heavy rainfall and hail storm. The provision of partial shade (30 to 50%) depends on plant species. Usually, partial shade is not needed after completion of germination of seeds. Prolonged shading may not be much useful for making seedling hardy.

Seedlings may also require prevention from diseases, the most cornmon are due to certain fungi. Suitable fungicides like copper oxychloride at 0.3 per cent for collar-rot, organic mercurial fungicide at 0.1 per cent for *Fusarium* spp., Dithane M.45 at 0.25 per cent for leaf spots, is sprayed or dusted ancVor drenched as and when disease appeare. Diseases like collar-rot is favoured by excess watering whereas *Fusarium* wilt is favoured by inadequate watering. Therefore, both an excess of watering as well as deficient watering should be avoided.

Seedlings should be raised under protected nursery where animals are not browsed. A wall or hedge may be useful to protect seedlings. A hedge which is hardy, thorny and browse resistant species like *Pithecellobiurn dulce or Prosopis juliflora* will serve the purpose.

Interculture and Weeding

The seedlings of plant species which require to remain 8-10 months or so in the nursery bed need to have once or twice shallow cultivation and more so weeding. The weed growth suppresses seedling growth. Therefore, it is essential, to do weeding wherever, weeds are proving harmful to the seedlings. Thinning is done if seedlings are crowded.

Shifting

Shifting operation in nursery stage is done in two conditions such as when (i) the seedlings are raised in polythene bags and (ii) shifting of seedlings of some plant species is done whose tap roots grow very fast and these young seedlings are shifted from primary nursery to secondary nursery. While shifting, the protruding roots are cut. Shifting becomes necessary with the seedlings raised in polythene bags, at monthly intervals. Shifting of such seedlings can also be done into earthen ware pots. However, shifting of seedlings to third place is rarely done. By shifting operations during nursery stage, seedlings become strong and stocky.

Pricking

Means taking out very young seedlings (3-4 cm) from primary nursery after completion of germination and planting out them into the other nursery bed for better survival and growth prior to finally transplanting in the permanent sites. Seedlings that have developed long tap root and profuse lateral roots are more prone to pricking damage than young seedlings with tap roots having no laterals. The best time of pricking is when laterals have just started developing. At this stage, the planting shock will be minimum and the seedlings are likely to produce healthy root system. Pricking damages are minimum in seedlings raised in polythene bags. Pricking time of some species is given below.

Time of pricking of a few species after sowing (in days)

Species	Days after sowing
Acacia auriculiformis	30
Ailanthus altissima	30
Adina cordifolia	50-60
Albizia chinensis	20
Albizia lebbeck	10
Alnus nepalensis	90
Azadirachta indica	50-60
Bombax cieba	45-50
Casuarinas equisetifolia	30-60
Celtis australis	45-50
Eucalyptus tereticornis	20-30
Grevillea robusta	45
Melia azedarach	25-35
Pinus roxburghii	45-50

Thinning

In the nursery bed, crowding of seedlings is usually seen. This condition is undesirable for producing healthy seedlings. Therefore, it becomes essential to reduce crowding of seedlings, which is achieved by thinning. It is an operation in which additional seedlings are removed in order to provide proper space to remaining seedlings for improving their growth.

Thinning is performed at young stage, delay causes seedlings to become week, feeble and lanky due to competition for light, soil moisture and nutrients.

Thinning in nursery bed is performed by pulling out the seedlings with hands. After completion of thinning light irrigation is given.

❑❑❑

Chapter-18

Propagation

Plants are raised by two ways:

i) By seeds

ii) By vegetative parts of the plant

Propagation by Seeds

Plants are raised either by sowing seeds directly at permanent site (*in situ*) or seedlings are raised in the nursery-beds or in any suitable container when they attain plantable size, are transplanted carefully at permanent site.

Direct sowing

Any artificial plantation can be raised either by sowing seeds directly at plantation site or in the nursery for raising seedlings, and when in nursery they attain plantable size, are planted at permanent site.

There are certain advantages and disadvantages of direct sowing of seeds at the permanent sites.

Adavnteges of direct sowing

- There is no need of nursery.
- Saving of time, man and material.
- Plant grow freely without any disturbances particularly in root system which occurs when seedlings are raised in nursery and uprooted for planting at permanent site.

Disadvantages of direct sowing

- More seed is needed.
- At permanent site effective supervision is not possible for large area.
- Weeds may suppress the newly growing plants; weeding is costly at permanent site due to large area covered.
- Plants cannot be protected from natural calamities as they can be protected in nursery beds.
- At permanent site, it becomes very difficult and rather impractical to provide most congenial soil conditions for seed germination, but it is easy to do so in nursery beds.

The following tree species are successfully raised by direct sowing of seed:

Acacia albida
Acacia auriculiformis
Acacia cyclops
Acacia mearnsii
Acacia nilotica
Acacia
Acacia seyal
Acacia tortilis
Ailanthus excelsa
Albizia lebbeck
Alnus nepalensis
Artocarpus chaplash
Azadirachta indica
Bauhinia variegate
Calliandra calothyrsus
Parkinsonia aculeate
Pithecellobium dulce
Prosopis alba
Prosopis chilensis
Prosopis cineraria
Prosopis juliflora
Salvadora oleoides
Cajanus cajan (shrub)
Cassia fistula
Cassia siamea
Ceiba pentandra
Cordial dichotoma
Dalbergia sissoo
Dendrocalamus strictus
Gliricidia sepium
Gmelina arborea
Grevillea robusta
Grewia optiva
Leucaena leucocephala
Madhuca longifolia
Melia azedarach
Moringa oleifera
Sesbania bispinosa
Sesbania grandiflora
Syzgium cumini
Tamarindus indica
Terminalia bellerica
Terminalia catappa
Zizyphus nummularia

By Raising Seedings for Transplanting

The following tree species should not be raised by direct sowing of the seeds at permanent site:

Acacia salinga
Adhatoda vasica
Anogeissus latifolia
Casuarinas equisetifolia
Celtis australis
Emblica officinalis
Eucalyptus spp.
Morus serrata
Muntingia calabura
Robinia pseudacacia
Tamarix aphylla
Trema orientalis

The following tree species should preferably be raised by transplanting of seedlings:

Acacia tortilis
Acacia salinga
Alnus nepalensis
Artocarpus heterophyllus
Azadirachta indica
Bauhinia variegata
Caasuarina equisetifolia
Celtis australis
Citrus aurantifolia
Dalbrgia sissoo
Emblica officinalis
Leucaena leucocephala
Prosopis cineraria
Prosopis juliflora
Quercus leucotrichophora
Robinia pseudacacia
Sesbania grandiflora
Syzygium cumini
Tamarindus indica
Terminalia arjuna
Zizyphus mauritiana

In a few plant species, entire seedlings do not show much success but when such seedlings are made into stumps by trimming 1/3 root system and leaving 2-3 cm shoot portion. Sometimes, over aged seedlings are also made into stumps for higher success. Stumps are completely devoid of leaves therefore transpiration from stump is reduced to minimum compared to entire seedling.

The following plant species may be raised by their stumps or root and shoot cuttings.

Albizia lebbeck
Albizia procera
Artocarpus chaplash
Bauhinia purpurea
Bauhinia variegata
Butea monosperma
Madhuca longifolia
Melia azedarach
Morus alba
Parkinsonia aculeate
Pongamia pinnata
Prosopis juliflora

Cassia fistula
Dalbergia latifolia
Dalbergia sissoo
Gliricidia sepium
Gmelina arborea
Madhuca indica
Schleichera oleosa
Syzygium cumini
Tamarindus indica
Tectona grandis
Terminalia arjuna
Toona ciliate

On the basis of duration of rotation tree species may be grouped into three categories.

Few examples of plants species on the basis of their growing period

Growing period	Name of species
Short lived	*Acacia catechu, Acacia nilotica, Casuarina equisetifolia, Ecacalyptus species, Gliricidia sepium, Grewia optiva, Leucaena leucocephala, Moringa oleifera, Morus spp. Prosopis juliflora, Sesbania grandiflora, Thespesia populnea*
Medium long lived	*Acacia nilotica, Manikara achras, Aegle marmelos, Albizia species, Artocarpus heterophyllus, Cassia fistula, Cassia siamea Casuarina equisetifolia, Dalbergia latifolia Dalbergia sissoo, Embica officinalis, Mangifera indica, Psidium guajava, Syzgium cumini, Zizyphus mauritiana*
Long lived	*Albizia species, Azadirachta indica, Cocos nucifera, Madhuca indica, Madhuca longifolia, Polyalthia longifolia, Shorea robusta, Tamarindus indica, Tectona grandis, Terminalia arjuna*

Propagation by Vegetative means

Plant species which do not set seeds or their seeds or are not viable or viability lasts very soon or by nature they have the capability of making union when they are grafted, and also they are able to produce roots when they are layered or planted by their stem cuttings and a few by root cuttings. Fruit crops like mango is propagated by inarching, veneer grafting, stone or epicotyl grafting, softwood grafting etc. Ber is successfully propagated by buddinq whereas sapota is propagated commercially by grafting. The sapota is grafted on rayan root stock.

The following are some species which can be progagated by branch/stem cutting :

Ailanthus excelsa
Albizia lebbeck
Hibiscus schizopetalous
Hibiscus syriacus

Albizia procera
Acacia melanoxylon
Bauhinia acuminata
Bombax ceiba
Boswellia serrata
Boussonetia papyrifolia
Bougainvilleas
Callistemon lancoolatus
Ceiba pentandra
Celtis australis
Cestrum diurnum
Cestrum nocturnum
Crataeva nurvala
Dalbergia sissoo
Dodonaea viscosa
Erythrina indica
Erythrina indica
Ficus benghalensis
Ficus benjamina
Ficus carica
Ficus retusa
Ficus racemosa
Ficus religiosa
Gmelina arborea
Erythrina indica
Gardenia jasminoides
Gliricidia sepium
Hibiscus Collins
Hibiscus mutabilis
Hibiscus rosasinensis
Ixora species
Jacaranda mimosaefolia
Jasminum species
Justifcia gendarussa
Lantana camara
Lantana sellowiana
Lannea grandis
Lannea coromendalica
Melia azedarach
Moringa oleifera
Morus alba
Nerium indicum
Parkinsonia aculeate
Pongamia pinnata
Populus species
Punica granatum
Salix alba
Salix babylonica
Salix duphnoides
Salix elegens
Salix fragilis
Salix tetrasperma
Salix wallichiana
Sapindus mukorossi
Tabernaemontana coronaria
Tamarix articulate
Thespesia populnea
Vitex negundo

The success of rooting can be increased many fold if these species are propagated under intermittent mist conditions. Tissue culture technique is no doubt wonderful for multiplication of millions of plants from a single mother plant. This technique is not only rapid but produces disease free plants. The valuable plant species like mango, teak, sal, coconut are being tried under tissue culture, will be of immense value.

Enhancement in root formation

A large number of plant species are difficult-to-root and even fail-to-root by stem cuttings pose a problem in propagation. The success of propagation depends on percentage of cuttings regenerate roots, and this variation not only exists between species but also among cultivars of the same species. Further, rootability of a particular cultivar is influenced by several internal and external factors.

In the regeneration of roots the role of carbohydrate and nitrogen is significant. High content of carbohydrate and low nitrogen favours rooting. Whereas high content of nitrogen and low carbohydrate does not favour rooting. High content of nitrogen develops fast growing, soft and succulent shoots. In many subjects, however, high content of carbohydrate in cuttings is not associated with ease in rooting. Auxins induce rooting in cuttings which can be further accelerated with the addition of sucrose indicating that a proper balance between auxin and nutrition is necessary for rooting.

The natural regenerative capacity of a particular plant may be improved further through manipulation in the treatment of mother plant, sensitization of cuttings with certain root promoter; selection of planting material, planting environment etc.

Treatment of mother plant

The mother plants are treated to get juvenile shoots. It is an established fact that the ability of cuttings to form roots increaes with increasing age of the plant. The main source of juvenile shoots is nursery where young seedlings grow. But many times nursery plants are not available then old plants remain the only source. The juvenile shoots may be obtained from old and mature plants in the following ways:

- By pruning, the old and unwanted shoots are removed, and after a couple of weeks juvenile shoots are developed.
- By heading back the main stem is cut either above or at the surface of the ground depending upon the prevailing situations. The new shoots arise from or below the cut portion. When such shoots become bit hard, are used as propagating material.
- By stooling or mounding, in which main trunk of the plant is covered properly with soil. After a couple of week juvenile shoots are produced depending upon species in question.

- It has been reported that spraying of gibberellic acid on non-juvenile and mature plants results in reversion to the juvenile phase in a few shoots.

The mother plants should not be given nitrogenous fertilizer. Restriction in nitrogen application will reduce shoot growth and hence increase accumulation of carbohydrate. Sometimes, irrigation should also be withheld to slow down the shoot growth. The plants should be allowed to grow in lull sunlight because partial shade is congenial to shoot growth. The shoot growth further may be reduced by root pruning. The mother plants are treated with growth retardants (CCC) and ethylene releasing chemicals, cause cessation of growth and accumulation of carbohydrate/nitrogen ratio of the shoot. High phosphorus level enhances rooting on cuttings. Zinc has been found to induce rooting in grape and plum cuttings. Zinc is required for tryptophan production in plants, the tryptophan is an auxin precusor.

Selected shoots may be girdled. It is performed by removing a ring of bark and thus blocking the downward movement of carbohydrates that get accumulated at the point which served a base of the cuttings on girdling. It is not infact only an accumulation of carbohydrates but also auxins, cofactors and other substances.

The mother plants should also be kept in the mist chamber till shoots grow and developed. Cuttings from such misted shoots improve rooting than non-misted ones.

In many plant species rooting is induced with a treatment of darkness (etiolation) to the basal portion of a newly emerging shoot with a strip of black polythene and the terminaL part of the shoot is allowed to grow in light. The period of etiolation depends upon plant species. Generally, it varies 7 to 121 days. Cuttings retain higher Etiolated level of auxin at the site of etiolation, is apparently one of the important reasons why etiolated cuttings root better than non-etiolated ones.

Selection of cuttings

In mother plants, there are some luxuriantly growing shoots, some are soft and succulent. Such types of shoots contain high nitrogen content and low carbohydrates. Studies indicate that lateral shoots in which growth rate is slow and carbohydrates have accumulated, are more suitable for taking cuttings, than rapidly growing succulent shoots.

The maturity of wood play an important role in the success of cuttings. There are evidences that hardwood cuttings contain more carbohydrates and less nitrogen than the softwood cuttings. However, planting environment may alter rooting success up to a certain extent regardless of maturity of wood. For instance hardwood cuttings are much suitable for outdoor conditions whereas leafy semi-hard and softwood cuttings under intermittent mist chamber or polythene tent. It has been experienced that hardwood cuttings of many plant species fail-to-root but soft and semi-hardwood cuttings of the same species root to a high percentage.

The presence of leaves on cuttings induce better rooting than their absence on cuttings in a large number of plant species. Hence, to induce rooting in cuttings, younger leaves with or without apical bud should be retained, and cuttings should be planted under minimum possible fluctuating temperature and humidity.

Treatment of cuttings

Besides, the above operations, auxins play crucial role in directing processes of root formation in cuttings. Auxins not only induct rooting in a considerable number of cuttings but also produce number of roots and quality of rootstem and recommended period required for rooting. Owing to quality root system under the influence of auxins, the survival and establishment of rooted cuttings is very high. Apart from the auxins, there are other types of chemicals which also help in rooting viz., non-auxinic chemicals, phenolics, vitamins cytokinins, inhibitors like abscisic acid, and growth retardants, unsaturated gases like ethylene, acetylene etc.

It is also known that the presence of inhibitors prevent rooting in many plant species. In such cases cuttings are soaked in water preferably in running water that results in leaching of inhibitors which facilitate rooting. Root formation in grape cuttings is improved when cuttings are planted after soaking in water. It has also been reported that pear cuttings after centrifugation improve rooting probably because of elimination of inhibitors rather concentration of root promoters.

Planting environment

Planting environment plays an important role in the success of rooting in cuttings in that well drained porous rooting medium induces better and rapid rooting because it provides a suitable micro-climatic conditions (air- orygen, carbon dioxide, plant nutrients in some cases, etiolation etc.) near the rooting zone. Amongst different rooting media, coarse sand is known to be the best medium for a majority of plant species propagated by cuttings particularly under intermittent mist.

In general, high relative humidity and low temperature are basic need for favour of rooting to a large number of plant species. However, different plant species respond to different range of temperature and relative humidity. These conditions are not lasting for long in outdoor, and hence, under mist chamber where in addition to these, intermittent misting of water is regulated, making cuttings cool and turgid, and prolonging the life of cuttings to give them chance to initiate roots. Under such conditions the general recommendations for temperature range is 20-25 degree C and for relative humidity it is 75-80 per cent depending on rooting period of cuttings.

To sum up, the above paras highlight the possibilities of improving rooting in cuttings in many ways However, response to certain method and practice depends on plant species, season of propagation, planting environment, use of root promoters, extent of accuracy of the propagators etc. Further more, it has been experienced that even almost congenial atmospheric conditions are in the reach of cuttings, it has been observed that many plant species fail-to-root. The exact cause(s) of failure in regeneration of roots opens the base for further systematic studies.

Propagation of tree species by softwood cuttings with the help of IBA (2000 ppm) treatment under intermittent mist.

Speices	IBA 2000 ppm % rooting
Albizia lebbeck	50
Albizia lucida	40
Bauhinia variegate	50
Bixa orellana	85
Bombax ceiba	100
Brownea ariza	50
Brownea grandiceps	55
Callistemon lanceolatus	80
Calophyllum inophyllum	100

Cordial sebestena	70
Crataeva religiosa	95
Ficus elastic	100
Ficus krishane	100
Ficus roxburghii	100
Lagerstroemia indica	80
Lagerstroemia thorelli	45
Peltophorum ferrugineum	95
Pterospermum acerifolium	80
Saraca indica	46
Solanum macranthum	80
Sterculia alata	85
Thespesia populnea	90

Besides, there are a few species which produce root suckers for their multiplication. These are:

Acacia salinga
Ailanthus excelsa
Albizia lebbeck
Anogeissus latifolia
Anthocephalus chinensis
Dalbergia latifolia
Dalbergia sissoo
Garcenia indica
Garuga pinnata
Gmelina arborea
Melia azedarach
Millingtonia hortensis
Murraya koenigii
Prosopis juliflora
Salvadora oleoides
Santalum album
Sapicum sebiferum
Xylia xylocarpa

The following tree species are successfully propagated by budding and Air-layering techniques.

Aegala marmelos
Annona squamosa
Artocarpus heterophyllus
Cordial dichotoma
Emblica officinalis
Grewia subinaequalis
Morus spp.
Parkinsonia aculeata
Syzygium cumini
Tamarindus indica
Terminalia arjuna
Zizyphus mauritiana

The following tree species may be tried by air-layering for vegetative propagation so that their genetic constituents are restored and variability restricted like stem cutting.Favourable environment for root formation in layers is rainy season. Further, rooting can be accelerated with the use of growth regulators like IBA, NAA and Seradix. Ring/girdling or notching when done in advance to subjecting air-layering operation favours rooting.

Acacia catechu
Bauhinia vareigata
Betula utilis
Bombax ceiba
Cassia fistula
Celtis australis
Cryptomerica japonica
Cupressus sempervirens
Dalbergia sissoo
Delonix regia
Eucalyptus macarthuri
E. microtheca
E. sieberna
Ficus carica
Fraxinus floribunda
Juglans regia
Magnolia grandiflora
Picea abies
Picea pungens
Pinus caribaea
P. insularis
P. gerardiana
P. khasya
P. massoniana
P. patula
P. radiate
P. roxburghii
P. wallichiana
Prosopis cineraria
Quercus incana
Robinia pseudacacia
Shorea robusta
Taxodium distichyum
Terminalia arjuna
T. chebula
Thuja orientalis
Toona ciliata

❑❑❑

Chapter-19

Management Techniques in Agroforestry System

Effective and efficient management in agroforestry systems require in different phases. These are:

- Pre-planting phase
- Planting phase
- Establishment phase
- Silvicultural management phase
- Maturing phase
- Harvesting phase

Pre-planting Phase

The following operations are essential to complete before planting.

Selection of site

This includes mainly selection of sites for undertaking agroforestry systems. However, there is no need, to select the site when agroforestry system is developed where agricultural crops are already growing but selection of site is definitely important where it is to be started freshly, for this purpose, landuse management is the best criterion.

Soil preparation

The next step comes for soil preparation. Before the crop is sown and the area where the trees are to be planted should be adequately prepared to avoid undue damage to the crop. It becomes also essential to make some protection from browsing animals. For this, fencing is done either with protective hedge with live plants or fencing with barbed wire. *Lawsonia inermis, Pithecellobium dulce, Carissa carandas, Parkinsonia aculeata, Dodonaea viscosa, Duranta repens* etc., are suitable to plant for hedge. Hedge is properly maintained by regular pruning.

Digging of pits

Before commencement of planting of tree species, pits are dug on demarketed spots 45 to 60 cubic cm size depending on soil type and species. When soil is shallow and nutritionally poor, the size of the pit is increased to 75 to 100 cubic cm.

It will be better, if pits are burnt with dry grasses or they are dusted with Melathion powder on their innerwalls. Thereafter, pits are filled with a mixture of top soil + 1/3rd to 1/4 part of well decomposed FYM or compost or leaf mold. After filling the pits, water is poured so that the soil of the pit is settled down properly. Then pits are left for planting.

When pH of top soil is unfavourable for a species but the soil at lower depth is ideal. In such cases, special technique such as planting in deep pits or application of soil amendments around the plant can be adopted to ensure the survival, till roots reach the favourable soil zone.

Selection of crop combination

On the basis of salient features mentioned earlier with certain modifications, the tree species as well as agricultural crops, grasses and animal combinations are made. Planting of tree species is done in such away that the total area does not cover more than 4 per cent. However, comparative economics has to be seen. Area under agricultural crops may be reduced if profitability is very high from tree species. In fact area under tree species is increased where agricultural crops fail.

The trees are planted on vacant land which is usually infested with weeds. It is essential to remove all weeds and follow suitable control measures otherwise young saplings will have slow growth, and in many cases, mortality will be very high.

Planting Phase

Selected tree species are planted in pits dug well in advance at proper distance depending upon tree species and kind of intercrop combination, purpose of planting, duration of rotation of tree species etc.

Planting of most of the species is done during monsoon, preferably beginning of monsoon. It is followed in areas where monsoons are poor. Whereas in areas having very heavy and continuous rains for months together, planting is done in the period of post monsoon, in the month of September-October. During this period, there is lot of moisture available in the soil; atmospheric humidity also remains moderate, and receiving one or two showers are common.

Usually, planting is done in afternoon but on cloudy day, it can be done throughout the day.

Further, in areas where assured irrigation facilities are available, planting can be performed during spring season or even any time of the year.

However, extremes of cold and heat both are not suitable for evergreen plant species. It is advisable to take advantage of rest period of the plants which is commonly occurred in winters. Seedlings become leafless. Such seedlings should be transplanted at permanent sites, they survive and establish on higher percentage.

The following are a few plant species suitable for transplanting in winter:

- *Aesculus species*
- *Celtis australis*
- *Jatropha species*
- *Juglans regis*
- *Populus species*
- *Robinia pseudacacia*
- *Salix species*

Generally, seedlings/saplings are planted along with ball of earth because certain species fail to establish when they are planted bare rooted. A large number of species do not respond negative when their additional root system is reduced up to certain extent by pruning (i.e. it is just triming of long roots).

It has been experienced that planting of tall seedlings (2-3 m) of one to two years or so age is successful provided they are raised in big containers preferably gunny cloth bags. This also does not require uprooting of plants from the bags instead intact plants i.e. along with bags are planted that results in cent per cent survival. Not only this, the greatest advantage is that such plants are protected from browsing animals due to their tall height than the height of sheep, goat, cow etc. Such plants also do not require any staking.

Cloth bag does not affect plant growth at all. It rots within a season.In Maharashtra, planting of mango cv "Alphanso" is very common with metallic tin in which mango grafts of two to three years old are retained. The metallic tin rots within the first rainy season. No adverse effect on plant growth has been reported.

Sometimes, ball method of planting young tree species of 2 to 3 years or so is also adopted by Foresters. In this method young trees are taken out carefully by digging deep around the base leaving sufficient area so as to avoid drastic root pruning. It is tried to take out plant along with maximum possible root system. Planting is done on cloudy weather day or there is dizzling that will be of help in higher survival and establishment of transplants.

Trimed root system accommodates well in the pit and has better survival and establishment. The regeneration of roots is better in trimed root system than intact root system because in no-pruned or intact root system, the long roots are made coiled while planting; this situation is not favourable for generation of new roots.

At the time of planting, little foliage is allowed on seedlings/ saplings. This can be achieved by defoliation and triming of laterals. Reduction of foliage load results in reduction in transpiration rate that help in better survival of seedlings.

While planting, near the base of seedling/sapling, soil is properly pressed this will help in reducing air-pockets from root system and plant remains in errect position.

Just after planting, water is dripped near the base of each plant. Some species are benefited with staking, for them wooden stakes are inserted by the side of plant and tied with string.

In agroforestry system, tree species are planted mainly by two patterns i.e. line planting (planting is done on straight line at optimum distance) and square pattern of planting (distance between trees is the same along and between rows).The other patterns of planting do not suit well. These are rectangular, triangular and quincunx planting.

Following is the method of calculating number of plants that can be accommodated per hectare by different patterns of planting.

1. Line planting

$$\text{Number of plants} = \frac{100 \times 100}{\text{distance within row x distance between row}}$$

2. Square planting

$$\text{Number of plants} = \frac{100 \times 100}{\text{Square of the planting distances}}$$

3. Triangular planting

$$\text{Number of plants} = \frac{100 \times 100 \times 1.155}{Squareof\ plantingdis\tan ce(i.e.thesideof\ thetriangle)}$$

4. Quincunx planting

$$\text{Number of plants} = \frac{100 \times 100 \times 2}{Square\ of\ the\ side\ of\ the\ planting\ square}$$

Methods of calculating spacing for given number of plants per hectare.

1. Square planting, $\text{Spacing} = \frac{100 \times 100}{No.\ of\ plants\ per\ hectare}$

2. Rectangular planting,

$$\text{Spacing} = \frac{100 \times 100}{No.of\ plants \times spacingof\ plantsinro\ wsorspacing\ of\ rows\ /\ ha}$$

3. Triangular planting, $\text{Spacing} = \frac{100 \times 100 \times 1.155}{No.of\ plants\ /\ ha}$

4. Quincunx planting, $\text{Spacing} = \frac{2 \times 100 \times 100}{No.of\ plants\ /\ ha}$

Establishment Phase

This starts just after planting of tree species. In order to obtain higher percentage of survival and establishment, and further growth of tree species it is essential to protect them from the following:

- Insufficient soil moisture
- Excess of soil moisture
- Heavy weed infestation
- Browsing animals
- Extreme climatic conditions
- Root pruning (it meets with deep digging near the row of the tree species by ploughing to ensure no root damage). Always leave a metre distance either side of the row.

After 60 days or so all gaps are replanted so that continuity of row is gained.

Agricultural crops are grown in between two rows one metre away from the tree rows. The careful pruning of the tree species for leaf fodder and fuelwood help in maintaining them in proper shape and form and do not restrict the competition for light, and agricultural crops can be grown throughout the full life of the trees. Not only this, crop grazing can be done in between two rows but it is not advisable to allow animals during the early growth stage of the tree species.

In order to replenish nutrients, application of manure and fertilizers is always advisable for economic return. The quantity of manure and fertilizers depends on the crop combination in question. For enhancing production, crops should be free from weeds.

Silvicultural Management Phase

Thinning

It is an operation under which weak plants are removed in order to provide proper distance between two plants which results in maximum utilization of sunlight, space, moisture, nutrient with minimum competetion among them and between agricultural crops.

Objectives of thinning

- To provide growing space to each plant
- To remove dead, diseased plants preventing any spread of disease to healthy ones.
- To remove plants/trees which are crooked, forked, basal sweep, roughly branched etc.
- To provide poles and posts that will produce some intermediate financial return from sale of thinnings.

In agroforestry, early thinning is important as it provides firewood, poles and posts. Thinning at later stage of development is not advisable. Thinning maintains even growth of trees.

Thinning of tree species has three pronounced effects on remaining population:

- Lower natural mortality,
- Deeper crown, and
- Crown expansion. Thus, photosynthetic activities are increased that result in better growth.

The intensity of thinning depends on existing plant population, type of tree species, climatic conditions, soil type, and length of rotation.

There are three methods of thinning namely systematic thinning, selective thinning and crown thinning. It is also advisable to thin out side twigs when heavy demand for leaf fodder coincides. Leaf fodder is usually needed in mid-summer or winter. As regard time of thinning, it is done immediately after harvesting an agricultural crop or after heavy grazing to reduce the effect on agricultural production.

Pruning

Pruning is the most important silvicultural tool of the agroforestry. It is done to maintain growth, volume, height of the trees etc. "Pruning is an operation of providing a well balanced growth pattern by cutting or otherwise removing an unwanted portion of plant which makes proper shape, enhances regular bearing, improves health, vigour, yield and quality of fruits, foliage, and facilitates better operations, such as thinning, picking, control of insect pests, diseases etc."

Types of pruning

- Light pruning
- Medium pruning
- Severe pruning

The light pruning is done at early growth of the tree species, it is also particularly done lightly in those tree species which are shy to pruning operations. Medium pruning is advisable to those tree species that bear flower and fruits on current season growth e.g. phalsa, fig, ber etc. Severe pruning is done in old tree species to induce coppice to regenerate new crop.

In Australia and New Zealand, different methods of pruning developed with *Pinus radiata* by Chris Borough (of CSIRO), Francis Clarke and Nail Barr the following methods are :

i. Fixed lift pruning,
ii. Selective pruning
iii. Variable lift pruning
iv. Tip pruning

However, in India such types of studies are lacking particularly in agroforestry tree species. The degree of pruning i.e. intensity of pruning has only been studied properly in a few fruit crops like ber, phalsa, fig, pomegranate, grape etc. Owners of the trees prune them based on experience.

In agroforestnj system, basal branch pruning and selective branch pruning are most advisable and commonly adopted; these operations maintain optimum volume of the tree species which does not interfere with agricultural crops.

When tree species spread parallel to ground in order to enhance more shoots the practice is known as **bending**. When tree species is cut at the ground level leaving a small stump from which arise a number of shoots which facilitates production of large amount of foliage; the operation is known as **Coppicing**.

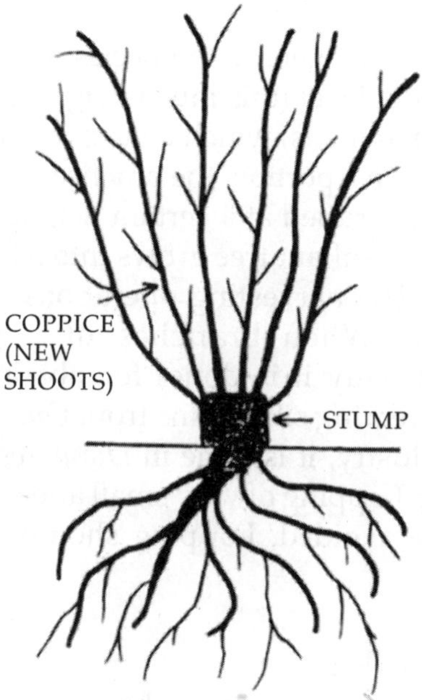

COPPICING

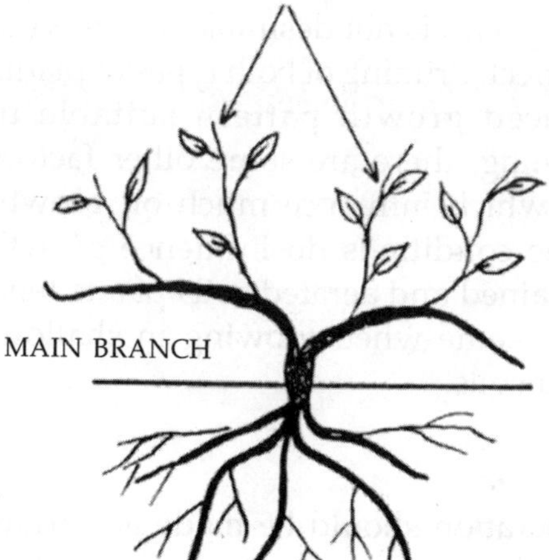

BENDING

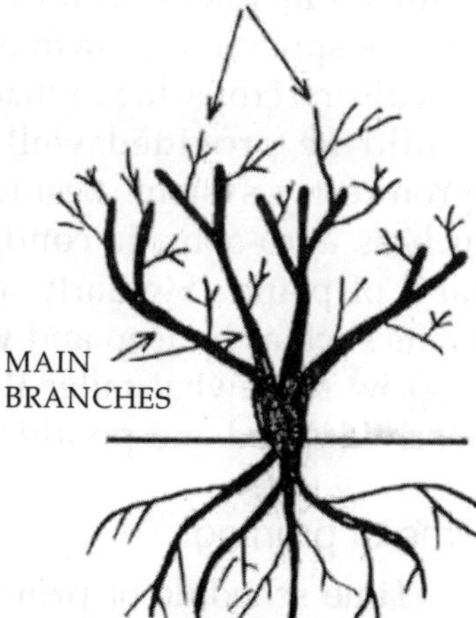

BUSHING

Figure Showing Pollarding, Coppicing, Bending and Bushing

Similar to coppicing, when tree species are cut 2-3 metres or so height from the ground level (this much height is beyond the reach of most of the browsing animals) which results in formation of a number of shoots from upper most portion, the practice is known as **pollarding**. When tree species are pruned at a certain height by allowing most of the healthy limbs the resultant tree grows into a bush, the operation is known as **bushing**. The harvesting operations become easy from a bush type of tree. When branches of a tree are prunded unsystematically (usually it is done) for obtaining leaf fodder for feeding animals, for rearing silkworms from *Querus incana, Morus spp.* etc. and for **bedi** industry, it is done in *Diospyros spp.;* the operation is known as **lopping**. Lopping of very small and young twigs, saplings and poles should be avoided. Lopping should be done at alternate season.

Factors affecting pruning

Growth habit: The tree species having their upright growth, should be pruned off by cutting their centre leader to induce more branches, that will also control height of the tree species, while other species which are spreading in their growth habit, should be pruned in such a manner that they may attain more or less upright growth because spreading growth of tree species is not desirable for growing agricultural crops due to shading effect. Pruning of both types of plants should be provided well balanced growth pattern suitable to agroforestry system. Besides pruning, there are some other factors such as agro-climatic conditions which influence much on growth habit of plants. Similarly, edaphic conditions do influence growth habits such as in deep and well drained and aerated soils plants tend to grow somewhat taller than the same when growing in shallow, poorly drained and poorly aerated soils.

Time of pruning

Time schedule of pruning operation should be made according to nature of tree species i.e. whether tree species is deciduous or evergreen, Deciduous trees require pruning in dormant period (trees shed their leaves in winter and go in for rest). During severe cold temperature falls to freezing, injury is likely to result in the case of pruned trees. Pruning in such cases should be delayed considerably.

Preferably, pruning of adult trees should be followed by young ones. Dwarfing effect is noted more in summer pruning than in winter pruning. In summer, the branches have attained functional status, and are helpful to the trees in the manufacture of adequate plant food material. Tree species pruned in late summer are also susceptible to winter injury. It is, therefore, essential to note that extreme cold and hot periods should be avoided while pruning a tree.

Bearing habit

Pruning should be done according to bearing habit of the tree with respect to fruits and/or foliage. Some tree species bear fruits and/or foliage heavily while others do not. For a better, high quality and profitable yield, an unwantea portion of the tree should be removed thoroughly.

Irrigation

Irrigation is not only required for survival and establishment of tree species but it become very essential throtghout life span of tree species though at wider interval but for agricultural crop it requires light and frequent interval for obtaining economic outturn. In fact, during rainy season, artificial irrigation is not required but when crops are raised during winter, spring and summer, irrigation is a must as and when it is needed. In order to economize water, if feasible, arrangement for drip irrigation may be done.

Weeding

Under agrisilviculture system intercultural operations keep down weed population. However, in heavily weed infested areas (infested with congress grass, *Cyperus rotundus* etc.) additional weedings may be required. Use of suitable weedicide can also be planned. In this way integrated methods of weed control will prove economical.

Plant protection

Infestation and infection of pests and diseases are inevitable in crops. Therefore, from the very beginning precautionary measures are advisable to adopt so that healthy crop yields are obtained.

Maturing Phase

It is commonly accepted that agricultural crops grow poorly between rows of mature trees. However, it always depends on tree species, tree size, pruned height and number of trees per unit area (stocking). Allowing annual grasses under mature trees for grazing cattle and other animal like sheep, goat, dear would be beneficial. Goat and sheep tend to clean up scrub, weeds and pruning and thinning debris better than other livestock and are therefore considered to have a good potential in older agroforestry systems.

Harvesting Phase

Depending upon age of the tree species and purposes, the harvesting is done. It is done in phased manner. When requirement arises for poles used in house construction, then trees are thinned at pole stage. This operation also offers good opportunity to near by plant with respect to space, nutrient, light etc. Always, harvesting is scheduled when high market prices are available. Later on, when trees attain adequate maturity they are completely cut at the ground level at onetime. This is done for the supply of valuable timber. Harvesting is done on the basis of diameter of the tree trunk i.e. at breast height (1.3 to 1.4 metres above the ground). Therefore, diameter at the breast height (dbh) is the best criterion. However, the best time to harvest depends on owner's preference. It is extremely important to have buyer ready before felling the trees because of the danger of degrading and the cost involved in storing the timber. The quality of the wood of many tree species deteriorate due to attack of fungi and many insects.

Harvesting of agricultural crops is done regularly according to their maturity stage. Organic refuse is turned down into the soil which maintain soil fertility. In pasture, livestock is allowed alternatively that can graze grasses as well tree species up to their height only.

❑❑❑

Chapter-20

Statistical Analysis

20.1. Statistics

Statistics are the numerical statements of facts capable of being analyzed and interpreted. Science of statistics is the study of principles and methods used in the collection, analysis and interpretation of numerical data in any sphere of enquiry.

20.2. Measures of Central tendency

Mean, median and mode are known as the measures of central tendency. The mean is sub-divided into following four types:

- Arithmetic mean
- Geometric mean
- Harmonic mean
- Weighted mean

20.2.1. Arithmetic mean

If $X_1, X_2, X_3, \ldots, X_n$ are the N observations, their mean X will be given by the following expression

$$X = \frac{1}{N}(X_1 + X_2 + X_3 + \ldots\ldots X_N) = \frac{\Sigma X}{N}$$

where,

SX = Sum of all the observations and

N = Number of observations.

20.2.2. Median

When all the observations are arranged in ascending or descending order of magnitude, the middle one is known as the median.

$$\text{Median} = \left(\frac{N+1}{2}\right)$$

20.2.3. Mode

That value of the variable which occurs most frequently or whose frequency is maximum.

20.2.4. Weighted mean (WM)

$$\text{WM} = \frac{w_1X_1+w_2X_2+w_3X_3+....+w_kX_k}{w_1+w_1+w_3+....+w_k}$$

$$\frac{\Sigma wX}{\Sigma w}$$

where, X is the value of an observation

and w is weight of X.

20.2.5. Geometric Mean (GM)

If X_1, X_2, X_3.. X_n are N individuals of a certain data, then their geometric mean,

$$=\text{GM}=\sqrt[N]{X_1.X_2.X_3...X_N}$$

20.2.6. Harmonic Mean (HM)

The harmonic mean of a number of quantities is the reciprocal of the arithmetic mean of their reciprocals.

$$\frac{1}{\text{HM}}=\frac{1}{\text{N}}\left[\frac{f_1}{X_1}+\frac{f_2}{X_2}+\frac{f_3}{X_3}+...+\frac{f_k}{X_k}\right]$$

$$=\frac{1}{\text{N}}\sum\left(\frac{f}{X}\right)$$

where $N=f_1+f_2+f_3+..f_k$

and f_1,f_2,f_3...refer to number of quantities.

20.3. Measures of Dispersion

20.3.1. Mean deviation

Mean deviation about the mean

$$=\frac{1}{N}S\ [x]$$

where$x$ = Deviation from the mean

[x]=Absolute deviation

N=Number of observations.

In case of frequency distribution –

Mean deviation about the mean

$$=\frac{1}{N}\sum[fx]$$

f=frequency of x.

20.3.2. Standard deviation (SD)

$$SD = \sqrt{\frac{\Sigma(X-X)^2}{N-1}}$$

In case of frequency distribution:

$$\text{Estimate of SD} = \sqrt{\frac{\Sigma(X-X)^2}{N-1}}$$

20.3.3. Standard error

The standard deviation of the sampling distribution of an estimate is known as the standard error of the estimate:

$$\text{S.E. of X} = \frac{\sigma}{\sqrt{N}}$$

20.4. Statistical Hypothesis

A mathematical statement is known as statistical hypothesis.

20.4.1. Null hypothesis

A hypothesis of no difference is called null hypothesis.

20.5. Test of Significance

Statistical procedure for deciding whether the difference under study is significant or non-significant, is called the test of significance.

20.6. Level of Significance

The maximum probability of rejecting the hypothesis when it is true, or in other words, the maximum probability of Type I error is known as "level of significance.

20.7. Degree of Freedom (D.F.)

Number of items or cell frequencies whose values can be determined at will

d.f. = N – K

where, N is the number of cell frequencies, and

K is the number of constraints.

20.8. Critical Difference

Critical difference may be defined as that least significant difference, equal to or greater than which all the differences are significant.

20.9. Block

A group of experimental units in which each treatment occurs the same number of times.

20.10. Coefficient of Variation

The estimated standard deviation expressed as percentage of the mean :

$$C.V. = \frac{SD}{X} \times 100$$

where, SD is the standard deviation and X is the mean.

20.11. Comparison or Contrast

It is defined as a linear function of the observations such that the sum of their coefficients is zero.

20.12. Correlation

The tendency of two variables to be related in a definite manner. For example, fertilizer and yield of crops would usually be positively correlated upto a certain level.

20.13. Correction Factor

A number subtracted from the sum of squares of a group of items to obtain the sum of squares of the deviation of each item from their mean.

$$\text{C.F.} = \frac{(\Sigma X)^2}{n}$$

20.14. Deviation

The difference between two values.Usually the difference between an individual variate and the mean.

20.15. Types of Distribution

20.15.1. Bionomial distribution

A distribution in which the number of successes or failures of a given event are recorded rather than some measurement.

20.15.2. Normal distribution

A bel shaped symmetrical distribution characteristic of many sets of data, in which 95% of the population lies within ± 1.96 from the mean.

20.15.3. Poisson distribution

A type of frequency distribution often associated with counts, especially when the event being counted is relatively rare in occurrence.

20.16. Error

The variation among variates that cannot be accounted for by the variation due to treatments, blocks and other factors controlled in the experiment.

20.17. Experiment

A systematic procedure for making observations under controlled conditions in such a way that they can be used for arriving at general conclusions regarding the population under study. When an experiment is designed to test two or more factors simultaneously, it becomes a factorial experiment.

20.18. Interaction

The tendency of one factor to have a different effect at different levels of another factor.

20.18.1. Order of interaction

Interaction between the two factors is known as the interaction of the first order. Interaction between n+1 factors is nth order interaction.

20.19. Population

Any set of individuals having some characteristic(s) in common.

20.20. Treatment

A single state of some factor being varied in an experiment such as rate of fertilizer, a variety of plant etc.

20.21. Variance

The sum of squares of deviations from the mean divided by the degree of freedom.

20.21.1. Analysis of variance

Analysis of variance is splitting up the total variation into various factors responsible for variation in a given population.

20.22. Regression

Defined as the increase in the dependent variable for a unit increase in the independent one.

20.23. Basic Principles of Field Experimentation

20.23.1 Replication

The repetition of the same treatment so that it appears two or more times in an experiment.

20.23.2 Randomization

The allocation of the treatments to the different plots by a random process is known as randomization of treatments. In this process each treatment has an equal chance to be allocated to an experimental unit.

20.23.3 Local control

The principle of making use of greater homogenity in groups of experimental units for reducing the experimental error is known as local control. This is done by dividing the experimental material (e.g., field) into homogenous groups of experimental units (e.g., blocks).

20.24. Standard Error of Mean (SE_m)

It is standard deviation of population:

$$SE_n = \frac{S}{\sqrt{n}} or \sqrt{\frac{S^2}{n}}$$

where, S is standard deviation and n is number of observations.

20.25. Standard Error of Difference (SE_d)

If replication number s variable for each treatment:

$$SE_d = \sqrt{\frac{S_1^2}{n_1} + \frac{S_2^2}{n_2}}$$

When replication are same:

$$SE_d = \sqrt{\frac{2S^2}{n}} \text{ or } \sqrt{\frac{2S^2}{r}}$$

where, S^2 = variance and r = replication.

20.26 F-Ratio

Ratio of two independent estimates of same variance.

20.27 Designs for Field Plot Experimentation

20.27.1 Completely randomized design (CRD)

The completely randomized design is most appropriate for experiments with homogeneous experimental units. It is commonly used in laboratory experiments where environmental effects are easily controlled. This design is rarely used in field experiments.

Analysis of variance

Source of variation	d.f.
Treatment (t)	t-1
Experimental error	(N–1) - (t–1)
Total	N–1

20.27.2 Randomized block design (RBD)

It is one of the most widely used experimental designs in agricultural research. It is characterized by block of equal size, each of which contains a complete set of all treatments. The chief advantage of design is that it reduces experimental error (through proper blocking) while retaining much of the flexibility and simplicity of the completely randomized design.

The split up of d.f. for a RBD is given below:

Source	d.f.
Replication	r–1
Treatment	t–1
Error	(r-1) (t-1)

$$\text{L.S.D.} = \sqrt{\frac{2EM_{88}}{r} \times t.05}$$

where,

EM_{88} = Error mean sum of square

r = number of replications.

Calculation of missing value in RBD

$$x = \frac{tT + rR - G}{(t-1)(r-1)}$$

where,

x = estimate of missing value

t = Number of treatments

T = Total of that treatment having missing value

r = Number of replications

R = That very replication total in which value is missing

G = Grand total.

After calculating missing value, SE_d is calculated as follows:

$$SE_d = \sqrt{S^2\left[2/r + \frac{t}{r(r-1)(t-1)}\right]}$$

where,

S^2 = error mean square

t = Number of treatements, and

r = Number of replications.

LSD = SE_d x terror

d. f.

Note: At the time of analysis reduce one d. f. from error d. f. because of missing value.

20.27.3. Split plot design (SPD)

The design developed specifically for factorial experiments is the split plot design. This design's basic feature is the use of two plot sizes one each for two variables under study. The larger plot size is called the main plot. The number of main plots in a replication corresponds to the levels of the first factor. Each main plot is subdivided into smaller size plots called subplots, such that the number of subplots in each main plot corresponds to the levels of second factor.

Using two plot sizes results in two levels of precision, one for main plot and one for subplot. Smaller difierences can be detected among subplot treatments than among the main plot treatments.

Split-plot design is used when

- The practical limit for plot size is much larger in one factor compared with the other.
- Greater precision is desired in one factor relative to other.

Analysis of variance in SPD : An example

Nitrogen levels = n

Varieties = v

Replication = r

Source d. f.

Replication	r-1
N (Nitrogen)	n-1
Error (a)	(r-1) (n-1) = E_a
(V) Varieties	v-1
N x V	(n-1) (v-1)
Error (b)	n(n-1) (v-1) = E_b

SE_d for N = $\sqrt{\frac{2E_a \; MS}{r \times v}}$

LSD for N = SE_d x tva d. f.

where,

E_a MS = Error a mean square

E_b MS = Error b mean square

SE_d for V = $\sqrt{\frac{2E_a \; MS}{r \times n}}$

LSD (V) = SE_dx tE_b d. f.

SE_d for N x V = $\sqrt{\frac{2E_b \; MS}{r}}$

LSD for N x V = SE_d x tvb d. f.

LDS to compare

Two variety (b) means at the same level of nitrogen (a)

$$\text{L.S.D.} = \sqrt{2E_b MS} \times t \;\; E_b \;\; d. \;\; f.$$

Two 'a' means at the same level of 'b'

$$\text{L.S.D.} = \sqrt{\frac{2 \;\; (B-1) \;\; E_b + E_a}{r \times v} \times \; tE \; \frac{a+b}{2} \;\; d. \;\; f.}$$

$\text{E} \; \frac{a+b}{2}$ = Weighted average of d. f. of a and b.

Calculation of missing value in SPD

$$x = \frac{pR + qM - P}{(p-1)\ (q-1)}$$

where,

x = missing value

p = number of subplots

q = number of replications

P = Total of known yield of whole plot treatment that contains missing subplot

R = Total of known yield of subplot treatment that contains missing subplot

M = Total of known yield of specific whole plot that contains missing value.

20.27.4. Latin Square Design (LSD)

Latin square design has capacity to block an area along two directions instead of only one. The two directional blocking commonly referred to as rows and columns is accomplished by arranging the treatments so that every treatment occurs only once in each row and once in each column. Thus, the Latin square design is especially adapted for use in an experimental site where there are two fertility gradients running perpendicular to each other. At the same time this design requires that the number of replications must equal the number of treatments.

Analysis of variance

Source of variation	d. f.
Squares	r-1
Treatment	n-1
Squares x Treatments	(r-1) (n-1)
Rows within squares	r(n-1)
Columns within squares	r(n-1) (n-2)
Total	$r(n^2-1)$

$$SED = \sqrt{\frac{2\ EMS}{r}}$$

C.D. = SED x terror d. f.

Calculation of missing value in LSD

$$X = \frac{t(R+C+T) - 2G}{(t-1)\ (t-2)}$$

where,

X - Estimate of missing value

R -Total of the observed values of the row that contains the missing value

C -Total of the observed values of the column that contains the missing value

T - Total of the observed values of the treatment that contains the missing value and

G - Grand total of the observed values.

20.27.5 Split-Split-Plot Design

To accommodate a third factor, each sub-plot in basic spilt plot design is further divided into sub-subplots and the randomization of the last factor is made separately for each sub-plot. The precision of comparisons among the sub-subplot treatments is the best, followed by those among the subplot treatments, with those among the main plot treatments having the least precision.

Analysis of variance

Source of variation	*Degrees of freedom*
Block	r-1
Nitrogen (N)	n-1
Error (a)	(r-1) (n-1)
Management (M)	m-1
N´M	(n-1) (m-1)
Error (b)	n(r-1) (m-1)
Variety (V)	v-1
N´V	(n-1) (v-1)
M´V	(m-1) (v-1)
N´M´V	(n-1) (.1) (v-1)
Error (C)	nm(r-1) v-1)
Total	nmrv-1

20.28. Factorial Concept

In this approach we compare all possible combinations of different levels of all the factors in one and the same experiment to test whether the factors actually interact with each other or are independent in their effects. Factorial experiments in which block effects are confounded with one or more interactions are called confounding experiments. An interaction can be completely or partially confounded with between block ariation.

20.28.1 Simple effects, main effects and interactions

An example,

Consider an experiment with two factors each at two levels -2^2 factorial experiment.

Factors : v_1 levels : n_0

v_2 n_1

Treatments : v_1n_0, v_1n_1, v_2n_0, v_2n_1

Replications : 5

Mean yield (kg)

Varieties	Nitrogen		Mean	Response
	n_0	n_1		$(n_1\text{-}n_0)$
v1	30	38	34	+8
v2	35	45	40	+10
Mean	32.5	41.5		9
$v_2\text{-}v_1$	+5	+7	6	+2

- The application of N has increased yield 8 kg in case of variety v_1 and 10 kg in case of variety v_2. These figures are known as simple effects of nitrogen.
- The simple effects of varieties v_1 and v_2 are 5 kg and 7 kg.
- In ease of n_0 and n_1 we observe that the average response of N is equal to 9 kg. This is known as main effect of nitrogen.
- The main effect of variety is 6 kg.
- Interaction of nitrogen with varieties (NV).

NV=1/2 (simple effect of ii with v_2 – simple effect of n with v_1)

For n factors each at 2 levels.

$$\text{Standard error of any effect} = \sqrt{\frac{2\ VE}{r_2^{n-1}}}$$

where,

VE is error variance,

r is the number of replications and

n is the number of observations.

e.g., 3 factors each at 2 levels

$$\text{S.E.} = \sqrt{\frac{2\ VE}{4r}}$$

20.29. Transformation of Data

Data transformation techniques are usually et1ictive in reducing heterogenity of variance when the variance is related to the mean.

20.29.1. Logarithmic transformation

The logarithmic transformation is most appropriate for data with standard deviation proportional to the means or when multiplicative rather than additive effects hold. These conditions are generally found with data that are whole numbers and cover a wide range.

Where zero values and very small values occur add 1 to each figure before taking logarithm. That is log (X + 1) is used instead of log X.

20.29.2. Square root transformation

Square root transformation is appropriate for data consisting of small whole numbers, i.e., data obtained in counting rare events, such as number of infested plants in a plot and weeds per plot. For these data, the variance tends to he proportional to mean. A correction may be made by taking the square root of each observation before proceeding with the construction of analysis of variance in usual manner. Square root transformation is also appropriate for percentage data where the range is between 0 and 30 per cent.

If most of the values are small, especially when zeroes are present, use $\sqrt{X} + ½$ instead of $\sqrt{X}$.

20.29.3. Arcsin transformation

The arcsin transformation is appropriate for data on proportions, either expressed as decimal fractions or percentages. The machanic of transformations are greatly facilitated by using a table of arcsin transformation. For values of 0%, substitute l/(4n) and for values of 100% substitute [100 - l/(4n)] where n is the number of units upon which the percentage data was based, i.e., the denominator used in computing the percentages.

□□□

[illegible] equally separate when errors are [illegible] [illegible] applied [illegible]

[illegible] transformation

[illegible] transformation is appropriate for data on proportions, [illegible] [illegible] frequencies or percentages [illegible] percent.

□□□